Winner of the Striding Edge
the Year, 2016.

'*The Gathering Tide* creates its ov
of one of England's last and richest wilderness areas – Morecambe Bay. The writing in her cycle of stories about humans and nature is full of earthy realism, authentic observation and quiet lyricism. It is a hugely impressive debut.' Mark Cocker

'Exquisite descriptions … with a forensic attention to detail that sparkles with lyrical imagery, mapping the history of the shores and sands.' Miriam Darlington, *BBC Wildlife*

'Evocative, muscular … an artist's eye for colour and a dynamic way with verbs.' Kathleen Jamie

'Undeniably beautiful writing … Lloyd's lyrical prose is smooth, like a pebble softened by the tide … A tribute by a gifted writer to the place she calls home.' Kate Feld, *Caught by the River*

'This poetic book is a map, a layered account of Morecambe Bay, its birds, rivers, names, bones, weathers, ghosts, tides and lives, its dangerous beauty, its tragedies … In mapping a place and sharing its beauty with her reader, she arrives at her own sense of belonging.' Gillian Clarke

'A vivid book with a landscape at its heart, redolent with the tang of original imagery. The hallmarks of good nature writing are in place – a seeing eye, that willingness to watch alone that deepens the bond between nature and writer, and the capacity for celebrating what Hazlitt called "the involuntary impression of things upon the mind."' Jim Crumley

'Slides effortlessly from the environment to history, to stories of other people, to personal anecdote … succeeds magnificently.' Robin Lloyd-Jones, author of *The Sunlit Summit*

Selected by Katharine Norbury as an *Observer* Book of the Year.

Also by Karen Lloyd

NATURE WRITING

Abundance: Nature in Recovery

The Blackbird Diaries: A Year with Wildlife

POETRY

Self Portrait as Ornithologist

AS EDITOR

North Country: An Anthology of Landscape and Nature

The Gathering Tide

A Journey around
the Edgelands of
Morecambe Bay

Karen Lloyd

Saraband

Published by Saraband
3 Clairmont Gardens,
Glasgow, G3 7LW
www.saraband.net

ISBN: 9781913393809

Printed and bound in Great Britain by Clays Ltd, Elcograf S.p.A.

10 9 8 7 6 5 4 3 2 1

For Steve, Callum and Fergus –
the three men at home

I look; morning to night
I am never done with looking.
MARY OLIVER

Dedicated to the memory of
Brian Fereday

Contents

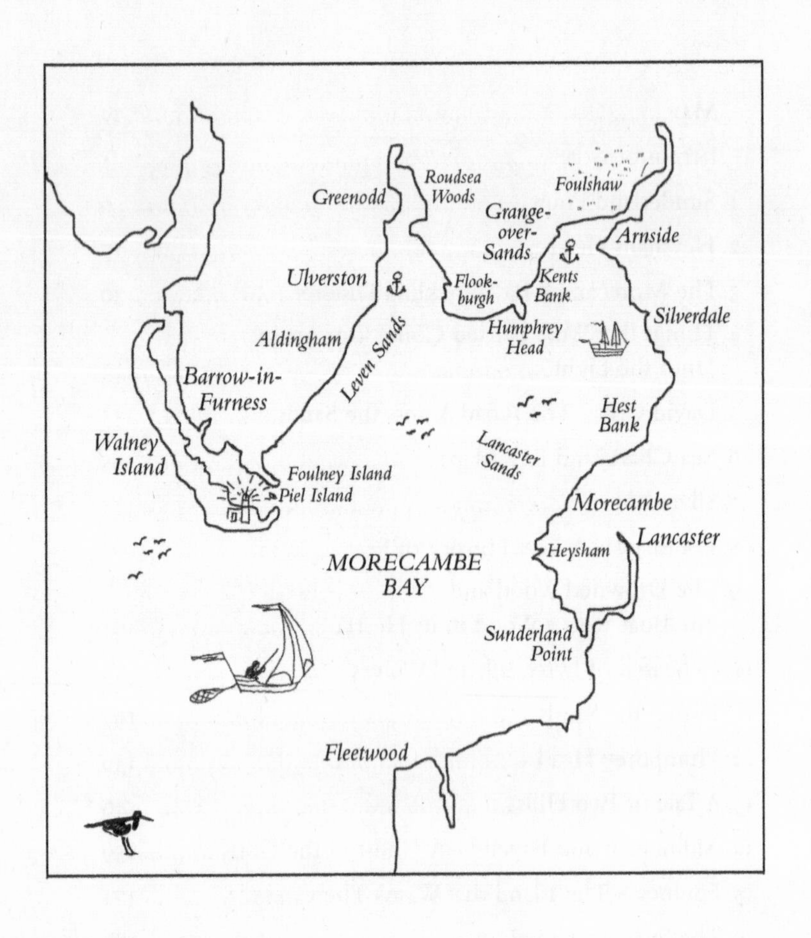
Roudsea
Woods
Foulshaw
Greenodd
Grange-
over-
Sands
Arnside
Ulverston
Flook-
burgh
Kents
Bank
Silverdale
Humphrey
Head
Aldingham
Leven Sands
Barrow-in-
Furness
Hest
Bank
Walney
Island
Lancaster
Sands
Foulney Island
Piel Island
Morecambe
Lancaster
Heysham
MORECAMBE
BAY
Sunderland
Point
Fleetwood

Introduction

It was one of those warm autumn days when I wished I'd left my jumper and coat in the car. I was walking at Silverdale on the Lancashire part of Morecambe Bay, along a wooded lane with the leaves overhead just on the turn, then through the kissing gate and onto Jack Scout. When I reached the Giant's Seat I stayed a while to take in the view, the tide ebbing and remnant channels of sea sparkling on the sands. There were shore birds gathered along the water's edge, but why on Earth had I left the binoculars behind, my phone too? The truth is that I relished a short time away from all that tech. As I descended the cliff path – a place I'd walked countless times before – I noticed a couple and a dog passing on the sands below, and further south towards Jenny Brown's Point, another couple. I had a momentary frisson of doubt, knowing how treacherous the sands can be and remembering stories of folks getting themselves into trouble when all they'd done was go for a walk on a sunny day – like today. But the bay is glorious, and that spectacular prospect drew me on.

Reaching the sands I strode out, heading round towards the village. I'd sneaked this unplanned walk after a work meeting nearby – and why not, I'd justified to myself; who knew how many such days there'd be between then and the turn of the year? I splashed through a shallow channel, the sands firm underfoot, then glanced at my watch. I was immediately snagged by familiar guilt; there was work to do, and I turned towards the shore. Both couples had disappeared from view. To reach the rocky edge there was a shallow pool to negotiate, but only a few metres separated me from land.

Then, beneath my feet came the sensation that what I stood on was about to collapse. Gingerly, I stepped back onto firm ground. I lived to tell the tale, but it was a sobering moment.

Since *The Gathering Tide* was first published in 2015, we have lost Cedric Robinson, Queen's Guide to the Sands for fifty-six years, whose knowledge, experience and good health saw him leading walks across the bay well into his eighties. Cedric's death in 2021 came just months after his beloved wife Olive's. Announcing his death, the Queen's Guide's Trust said, "Now, Cedric, it is time to rest your sandy feet and keep an eye on us from up there." This makes it even more poignant for me to recall the day I spent with Cedric on the bay, checking the condition of the River Kent ahead of a walk the next day. It had been a similar day, the sky bright blue with huge assemblages of callous clouds all reflected, travelling in pools of seawater as Cedric drove the tractor miles out onto the sands. I wrote then how I'd spent years in the mountains, getting myself up and back again, and sometimes becoming disorientated in mists, "but this was something new, this sense of flat space.... I wasn't alone and of course I couldn't have been in safer hands, but there was something about this landscape and our remoteness from land...It was a new sensation for me." In 2020 The Queen's Guide Trust appointed fourth-generation Flookburgh fisherman Michael Wilson as the new guide, one of the very few people who is on intimate terms with the bay and all its foibles.

Since first publication, in so many ways the ground has been continually moving under our feet. Certainty – whatever that was – has been suspended. Nothing is as it was last year, last month or even last week. Covid happened. Brexit happened. Political chaos is everywhere. Climate chaos is happening. Biodiversity loss is happening. But still, here we are, picking ourselves up and setting off once more, if more trepidaciously, into the future. I read recently that one of the most important tenets required of us in all this uncertainty is rather than being beset by the rising tide of problems, we should feel energised and excited by the solutions. And it is indeed affirming to know that so many people – local and from further afield – care deeply about the bay, not just because it is a beautiful place, but because it provides home and sustenance to such a significant amount of wildlife. Take the birds. How many other places are there in Britain where you can see bearded tits, marsh harriers, avocets, geese, curlew, lapwing, oystercatcher and osprey? One of my students recently sent me a diagram illustrating how perfectly adapted wading birds are to this kind of rich ecosystem, each species' bill length designed to exploit a different layer in the silt and enabling them to find the invertebrates that dwell amongst the continuous turmoil of tide, sun, wind and rain.

We know that to be excited by (or even in love with) such natural landscapes is good for us. We go into nature and suddenly, the weight of the world lifts from our shoulders. This is what the bay offers – masses of space, masses of nature and beauty. Wordsworth got it right when recalling crossing the bay by carriage he wrote of it: "*...under a genial sun/With distant prospect among gleams of sky/And clouds and intermingling mountain tops/In one inseparable glory clad*."

Part of the beauty – or the gift, if you like – of Morecambe Bay, is that there's plenty of it. It's a place that offers space to be and space to think. When I set out to write the book, I envisaged a neat, year-long journey of walking and researching and thinking and writing – and rewriting, of course. But with over 60 miles of coastline to explore and with life sometimes getting in the way, the journey didn't take place in quite such shipshape fashion. The bay is a place I thought I knew well; I was brought up here and have lived most of my life within a stone's throw of the shore. Part of the joy in writing therefore was in finding unexpected corners or images or histories; stories I unearthed much in the way the tide re-reveals features and artefacts as it cycles through time. It was exciting, for example, to research the bay's pre-history through the archaeology of the Victorian era and more recent excavations in a series of caves, places where bones revealed the presence of wolves in the landscape and dolphins in the marine ecosystem. I studied David Cox's *Market Traders Crossing the Sands* – one of dozens of paintings the artist made of the bay, which reveal the mechanisms of the area's society, transport and economy. Studying ancient sea charts allowed me to discover how over centuries fully rigged sailing ships plied their way into the far reaches of the bay. What a sight that must have been!

One of the more recent seismic shifts that we have had to reconcile ourselves with – and rightly so – is the Black Lives Matter movement. *The Gathering Tide* begins with a visit to the grave of an unknown African male who may or may not have been a slave, but who came off a ship having taken ill and was then abandoned at Sunderland Point in the early 1700s. Days later he was dead. He was buried – or so the story goes – in a grave beside the sea. The grave is within sight of Heysham's nuclear power station, and if ever there was one, here is a place where time, technology, and an unfortunate manifestation of white privilege collide. Revising this introduction, I needed to reflect on how the grave site has subsequently been developed. It is no longer a "lonely hollow." Now there's a broad path and a permanent installation – a stone chamber – which feels to me to be unfortunate on so many disturbing levels. The development is part of a misguided plan to develop tourism and thereby the local economy. This might be bad enough, but even now – several years after the installation was built, the grave of the unknown African must be accessed by parading past the structure, as if this were the main event. Not only this, but many of the trees planted in plastic tubes along the site of the chamber are dead, and those that remain alive now contort from their abandoned, degenerating plastic tree guards. As for the grave itself, it is still marked with the term (for it is a term, and not a name) "Sambo's Grave". Perhaps there's an element of false-memory syndrome at work here. It's as though by looking primarily through the lens of tourism, it's all ok, just something we call "heritage", as if that is somehow separate from lived experience. But memorialisation demands our fuller attention, together with an admission of guilt and the ability to say we're sorry. We cannot roll back time, but if we are to truly to demonstrate that we consider people of all races equal, and if Lancaster is to fully acknowledge and deal with its major role in the international slave trade, then here is an open opportunity to engage fully and meaningfully, and to make appropriate reparation.*

It's both heartening and humbling to see how *The Gathering Tide* has become embedded in the cultural and imaginative life of this extraordinary area. It is also gratifying to know that the book's journey continues through this new reprint. Life moves on, but books are here with us for generations, and I'm delighted to continue to share the memories and experiences that make up this literary ramble around the coast of Morecambe Bay. Mark Cocker described the bay as "one of England's last and richest wilderness areas," and this may indeed be so. Much of its fascination lies in how the bay manifests as an extraordinary fusing of land, sand, sea and wildlife, encompassed by the exquisite prospect of rolling hills, tapering headlands and that coronet of Lakeland mountains. But then I'm biased of course, because this is my place.

Karen Lloyd, Cumbria, November 2022

* My full length essay 'White on Black: the Ongoing Problematic Narrative of *****'s Grave' is published in *North Country: Anthology of Landscape and Nature* (Saraband, 2022).

One

Sunderland Point

I STOOD AT THE EDGE of the bay, looking out to where cloud shadows fell onto the immeasurable sands, colouring them deep Prussian blue and red ochre. The whole saltmarsh was silent, the kind of silence that hums in the ears, and it spread over the singing blue of a frozen morning. Then a redshank materialised from a channel close by and unfurled itself skywards, casting its singular ticking call into the sky. My cover was blown and the whole marsh knew I was there.

Over the fells of Furness and further, beyond the Duddon shores, Black Combe had turned other-worldly, snow-capped, like Mount Fuji transplanted. I'd driven through the brightening of an early January morning, the lanes funnelling into a narrow single track, passing caravan sites and low-lying winter fields where flocks of goldfinches sparked out of the hedgerows. There was a sense of getting closer to the coast; the way the light shifted, paling as it fell towards a seaborne horizon.

Then a place to park by the edge of the bay and, at the road's end, a few houses and gardens sheltered by trees. Rooks were high up in the motionless branches; feathers ruffling, wings restless. I pulled on boots and gloves, and gathered together the bits I'd need – binoculars, notebook and pen – and stuffed them into pockets before setting out to walk in extraordinary light. A pair of curlew flew up. As though choreographed, more appeared, surfacing and launching forward in sequence, out from the sea-pools and channels that threaded curving ribbons of water towards the bay, a good half-mile distant over the marsh. Arcing above me, they surged seaward, making the sky brimful with their bittersweet calls. This, of all the cadences of the natural world, fills me with the fizz of wild electricity. Oystercatchers evolved out of the distance, the flash and message of them, the sound building to a crescendo, insistent. Over the shoreline dunlin massed, flowing and *peep-peep*ing through the breath-cold morning sky. Underneath all this, the persistent chatter of turnstones.

Constellations of birds gathered on the sea margins, waiting for the returning tide. Curlew, redshanks, greenshanks, dunlin, knot, oystercatchers and, out on the empty sands, the gulls, splashed like distant white stars. Even further, milky white galaxies of birds worked, probing the sands for food.

A bank had been constructed as a flood defence, protection for the farmland on Sunderland Point. The line of it stretched all the way down to the end of this thin strip of land, the southern marker of Morecambe Bay. Its base was bordered with scoured marsh grass that'd been raked out and deposited by the sea, the line of it meshed together with knots of blue nylon rope, sea-washed plastic, fire-blackened wood, gull feathers, bottles and mussel shells. The tides had been full and I knew that the sea would come in this far again, covering the path.

As I walked, a pair of blackbirds bobbed ahead of me along a line of wavy fence posts tilted at angles and spliced together with rusted wire. Out of the far distance a crack of sound came, like gunshot, and countless birds on the sea-edge lifted into the sky, clouds of dunlin and knot moving in front of each other, white on black, and black on white.

It was nearing mid-morning and the sun was still low-slung in the sky. Sea-pools reflected the shift from yellow sky to cerulean blue. Vapour trails formed and dissolved. After months of rain the New Year brought the gift of intense cold weather, and I was glad of it; we need our light this far north, as much of it as we can get. A tractor had been along the path, creating ruts the depth of canyons, and the weather had created its own temporary geology of freezing, melting and re-forming inside the tracks. There were whorls of opaque ice, splintered white crystals, dark, depthless frozen pools and boot-marks that had frozen and been preserved. They reminded me of the Neolithic footprints exposed by shifting tides not all that far away, just down the Lancashire coast. The footprints of adults, children, cattle and even dogs; ephemeral archaeology, dug out by weather and water. I wondered about my footprints being exposed again and found in a couple of thousand years. What would they make of me?

I came to a stile set in a wall and stepped into a small, grassy enclosure bordered by drystone walls, a farm fence and winter-black gorse. Flat on the grass was a gravestone, with a small wooden cross at its head. A robin kept an eye as I walked towards it.

Here lies
Poor SAMBOO
A faithfull NEGRO
Who
(Attending his Master from the West Indies)
DIED on his Arrival at SUNDERLAND.

Full sixty Years the angry Winter's Wave
Has thundering dash'd this bleak & barren Shore
Since SAMBO's Head laid in this lonely GRAVE
Lies still & ne'er will hear their turmoil more

The grave has a life of its own; it seems to change with the seasons and the weather. I'd been here before with my husband and son, and on that autumn day there'd been: a collection of bright painted beach stones surrounding the cross, many with messages of love to the slave boy; a key ring with a small black and white dog attached to it; a painting of a fairy; a miniature bear and the carved wooden figure of an African antelope.

Going back there on that January morning, Christmas ephemera had been left: a pair of tiny angel figurines and a rusted metal coil that had been a miniature Christmas tree. There was a prayer book made of painted plaster with the message 'In loving memory', a plastic hedgehog and a child's plastic bangle dangling from the cross; its hollow gems refracted the winter light onto the pebbles at the base.

But there had been an unwelcome addition.

A slate panel had been hung from the horizontal bar so that most of the cross was now hidden from view, and the slate had become the first – no, the only – thing you'd see as you entered the little enclosure. In neat lettering someone had written the words of the hymn 'Amazing Grace', and added at the bottom, 'I come to this peaceful and beautiful place to give thanks and enjoy this charming and delightful special burial ground.' Myself, I liked the plastic hedgehog better; it had more eloquence, even for the simple nonsense of it.

* *

Reverend James Watson, a retired schoolteacher, installed the grave some sixty years after the boy's death. In 1796 he decided to mark the boy's death for posterity, and set about raising the necessary money for the grave and composed the eulogy inscribed on the stone. The story of the slave boy changes depending on where you read it. An article published in the 1822 edition of the *Lonsdale*

Magazine, a 'provincial repository', containing 'literary, scientific, and philosophical essays, original poetry, entertaining tales and anecdotes, miscellaneous intelligence etc,' suggests the boy died from his inability to speak English. Somehow though, I doubt it.

> *After she had discharged her cargo, he was placed at the inn … with the intention of remaining there on board wages till the vessel was ready to sail; but supposing himself to be deserted by the master, without being able, probably from his ignorance of the language, to ascertain the cause, he fell into a complete state of stupefaction, even to such a degree that he secreted himself in the loft on the brewhouses and stretching himself out at full length on the bare boards refused all sustenance. He continued in this state only a few days, when death terminated the sufferings of poor Samboo. As soon as Samboo's exit was known to the sailors who happened to be there, they excavated him in a grave in a lonely dell in a rabbit warren behind the village, within twenty yards of the sea shore, whither they conveyed his remains without either coffin or bier, being covered only with the clothes in which he died.*

Elsewhere it's suggested that the boy was the only survivor of a shipwreck off Sunderland Point, or died from a broken heart after his master left him. Some reports suggest that he might've died from a European disease to which he had no immunity. This seems more like it, makes more sense, putting two and two together. But anyway, I wanted to dig deeper.

* *

Sunderland's history is bonded tightly to Lancaster's. The village was developed as an 'outport' for Lancaster early in the 18th century by the Quaker industrialist Robert Lawson. Lancaster's economy boomed as a direct result of the slave trade, and the city's renowned Georgian architecture was built on the back of it. Lancaster became the fourth largest city in the UK involved in the slave trade, following only London, Liverpool and Bristol.

I'd been to visit Lancaster's Maritime Museum on the Georgian quayside, housed in the former Custom House, and I'd seen a troubling image. A pattern in black and white caught my eye from across one of the rooms. It reminded me of an African textile border print. But, walking closer, it dematerialised as pattern and became instead a kind of instruction manual, a graphic illustration showing the method of packing slaves onto ships. Four hundred and fifty people would have been crammed onto the four decks of a Lancaster slave ship. At the bulkhead end they were arranged at odd angles so that no space was wasted.

The picture stirred in me one of my most powerful memories from primary school. Studying the slave trade, we'd been shown a similar image and my young eyes had taken in all that crushed humanity. Between 1736 and 1807 Lancaster's ships carried around 29,000 slaves from West Africa to the West Indies.

The boy 'Sambo' would have been a servant, a cabin boy or a personal assistant. To have a black slave was considered a luxury, a sign of status, and whilst slaves did not come into England en masse, at the time of Sambo's arrival and death at Sunderland, around 40 country houses in the Lancaster area were known to have had black servants.

The city developed a different kind of slave ship, smaller than most others were at the time. Designed with a shallow draught, they could venture inland, navigating up rivers on the Windward Coast (modern Liberia and Ivory Coast), the River Gambia, and explore African estuaries where slavery was becoming a more specialist pursuit. They carried beads, bracelets, mirrors, hawk bells, clothing, hats and brass pans. They also carried iron ore, which was then a scarce resource in Africa. Ore had been dug out of the

Furness hills for centuries, and the bay area is also rich in limestone, used as flux in the smelting process. The remains of lime kilns across the area are testament to the extent of the industry. Iron was smelted into chunks here and exported along with all the other ephemera of slave-buying currencies.

The earliest recorded ship sailing from Sunderland to Jamaica was in 1687. Named, somewhat inappropriately, *The Lambe*, she set sail during the reign of King James II of England, VII of Scotland. Though this seems to be extraordinarily early, it shows the scope and aspiration of tradesmen from a small city in the north-west of England. I'd read about young men being plied with wine in the local inns, and once unable to stand upright they were carried off and put onto ships at Sunderland Point. I wonder which must have been worst – the mighty hangover, the rolling ship or the realisation that there was no going back? They must have thought they were sailing to the edge of the world.

In the Maritime Museum there's a poem from the time, a kind of gruesome justification for the slave trade.

I own I am saddened by the purchase of slaves
And fear those who buy them and sell them as knaves.
What I hear of their hardship, their torture and groans
Is almost enough to draw pity from stones.
I pity them greatly but I must be mum,
For how could we do without sugar and rum?

The Quakers were heavily involved in the Lancaster slave trade. It seems that they were able to turn a blind eye to the 'torture and groans'. It's unclear how they squared this with their belief that Christ was available to everyone, there for the taking; just the small matter of the colour of skin that meant you were in, or out.

* *

I wanted to see Sunderland Point in a different light. Six months later, with the tide far out on an early summer day, I cycled the tidal causeway road that links the village to the rest of the world. The first section of the road dives below the level of the marsh, so that my view of the river margin was seen through a pink bloom of sea thrift. Occasional cars ploughed the road, all but their roofs hidden from view.

High up, a splash of skylarks filled the sky, though with the light so intense it was impossible to pick them out. Their songs trickled down over the marsh. The ground was furrowed by water channels and streambeds, and water left by the tide oozed and prickled, wicking its way underneath the surface. In a deep muddy hollow a slither trail of evidence; something had been here, moving towards a stream and a hidden route to the river. I thought, otters.

Arriving in Sunderland village it was as if a warp in time as well as space had been crossed. Take away the streetlights and TV aerials and you could imagine yourself back in the 18th century. There was a small, walled orchard with grass so high it would part like the sea as you waded through it. Windows looked out onto the river. A small boat zipped out to sea and a line of geese came flying an oscillating line down-river, their call a wave pulse of sound. If I lived here, I'd never get anything done with all this to look at – the birds, the boats and the ships. I'd have to keep checking in, see what was happening.

There was a sign fixed to a gate: 'Local History Book for Sale. No obligation to buy. Come in and have a look'. I walked through the gate and into a walled garden where swallows dipped and dived on a pathway of air, chittering and swooping in and out of nests underneath a covered porch. The garden borders were filled with canna lilies, delphiniums, lupins. Inside the house a radio was playing, but no one answered my knock. I began to feel like an intruder. A robin landed a foot away from me on the garden wall, inspected me with his head cocked to one side before flitting away on a burr of quick wings. I watched the swallows for a few moments more then walked to the front door. On the way, a woman in an upstairs window caught my eye and waved.

She came to the front door and apologised for not hearing my knock. 'I'm in the middle of painting the bathroom,' she said, and I told her I was sorry to interrupt the flow. She fetched a copy of the book and we fell into conversation.

She told me that her husband was a fisherman, but that the fish don't come like they used to, and you couldn't make a living any more. Before I left she said, 'Come back again if you want to know any more.'

I thanked her and said, 'I'm off to the grave now.'

'So you know the way?' she asked.

'Yes, I've been before.'

'It's always changing,' she said. 'People bring things all year round.'

'I know. Last time I was here there were Christmas decorations, a miniature tree, that kind of thing.'

As I walked to the gate she said something else, and her words hung in the air after we said our goodbyes.

'That's if he *is* buried there.'

I left the bike, out of habit locking it to a fence, and walked across the isthmus of land to the grave site along a lane that was thick with June's offerings; elder flowers, wild rose, rabbits disappearing under hedges and a badger sett, recently swept clean, the way they do.

The grave had changed again. The too-big sign was still draped over the front, but someone had attached coloured ribbons to it.

Maybe one day someone will turn the slate over, start a petition: 'Let's stop calling him Sambo. How's about "Sam"?'

The Christmas tree had gone. There were new pebbles, newly painted too, one with the words, 'In memory of Sambo, God Bless from the Marsh History Society.' A boat constructed from tin, some origami, a plastic car and another pebble: 'Have a happy day Sambo, Kylie X'. The plastic hedgehog had gone. It'd been replaced by a frog.

For me there's a sense of tension around the memorial. There's a temptation to regard even the idea of the grave and its story as relics from the past, as anachronisms. But almost daily emails arrive asking me to sign petitions against slavery. There's even an ad on the TV – 'Help put an end to modern slavery' – and regular stuff in the papers too. It's almost 280 years since the boy was abandoned at Sunderland, but dig just a little and you find that there are likely to be more people bonded into slavery now than ever were during the Atlantic slave trade. Processing this, squaring it with the small grave in the field, is difficult.

* *

I walked back along the lane to the village, and stopped outside 'Upsteps Cottage', the place the slave boy had run to. He'd fled, so the story goes, up the flight of external stone steps and into the brewhouse storeroom, and it's here that he was left alone to die. But there's a mystery, a question. Some contemporary reports indicate that the boy's body was carried to a field behind the brewhouse and interred there in unconsecrated ground, rather than at the site of the grave. And why would you carry a body any further than you needed to?

By 1796, when the Reverend Watson decided to mark the lonely grave beside the sea, all was not well in either Sunderland or Lancaster. Ships were being built on a larger scale, and there were ongoing problems with the River Lune silting up. To facilitate easier access to the city Glasson Dock had been constructed just upriver. But it all came too late. With its larger and more accessible port, Liverpool was fast becoming the centre for trade and Sunderland had become known as 'Cape Famine'.

I remembered the words of the woman I'd bought the book from. 'That's if he *is* buried there,' she'd said, and I wondered if the boy really had been buried in the lonely field beside the sea. Or was the grave a creation, a kind of Turner Centre of the day designed to bring about regeneration? In the late 1700s the benefits of taking the sea air and sea-bathing were being promoted as aids to health. People had begun to visit Sunderland for recreation, and I wondered if this had got our Reverend thinking. A short walk across the fields may have been just what people needed, a focal point, or a reason to walk. After all, it's why I went.*

** See the introduction for an updated note on this chapter.*

Two

Heysham Head

At any time the graves are a remarkable sight. Six tombs hewn out of the rock on the top of a headland overlooking the bay, sarcophagus-like, and lipped for a covering of stone or wood. Less obvious at first, two more of them further back in the shelter of a line of wind-blasted trees. One infant-sized and, close by, another, the length of a child. In their own time they must have been eloquent, intended for highly regarded people, a family perhaps, who might continue to influence their community even after death. On the frozen January afternoon I went to see them, it was as though the bodies had morphed back into their stone graves. Rainwater, collected and frozen, gave them a sense of the corporeal again. Shrouds made of layers and whorls of ice were packed tight into each one.

⋆⋆

These were short days when afternoons segued into evening before you knew it. The light had intensified early and looking out from the beach at Morecambe Bay, brimful with the tide, the mountains of Cumbria appeared like a lost snowbound kingdom, coloured a ghostly, veiled pink. The beach at Half Moon Bay was alive with dogs that ran into the sea or bothered other dogs as their owners chatted with one another, idly throwing sticks or balls. A man walked past me and, whilst looking at his dog, uttered 'Nice day in't it,' and I answered yes, it really was. Two different kinds of black and white on the sand: oystercatchers and lapwings dotting about, following the imprint of the waves as they rose and fell. The light over the bay clarified all the time so that across on the long reach of the Furness peninsula individual houses and farm buildings were visible. A single glint of sunlight was reflected back from a window in the hills. At Barrow the land slid seawards and the giant hangars where nuclear submarines evolve mirrored the scale of the two brooding nuclear power stations half a mile south from where I stood. In the middle of that great expanse of water one small sailboat caught and reflected the light on its bright white sails.

As I began to walk, one ferry was embarking for the Isle of Man or Ireland and another was heading back into Heysham Port. I followed a path that meandered past gorse bushes and the skeletal heads of cow parsley, dogwood stems plum-red against the grass. Coming towards the top of the headland, the sun caught the ruins of St Patrick's Chapel. It seemed to glow, giving out a warm, ochre-red light. Through the Saxon arched doorway the sea of Morecambe Bay was a vivid blue. On the top of the ruins a pair of magpies fretted, casting out clattering cries like ill-tempered guardians. There were visitors, though very few; a woman sitting in the grass with her eyes closed and her face raised up to catch the slight warmth from the sun.

I looked down at the stone graves. They exude mystery, and enigma. There's nothing else like them in the country. I wanted to be able to place them accurately in their own time, but any grave goods are long gone and it's virtually impossible to locate them in a particular culture or to impose a date on them. It's thought they pre-date the first chapel to be built on the site. The 7th to the 11th century was a time of huge flux and change at Heysham that witnessed a mixing of cultures and belief systems. I wondered if the proximity of the graves to the chapel might only confuse things, or conflate the idea that they both existed as integral parts of an organised and influential early Christian site.

I wanted to understand something of their significance, but to do this I had to take away the present, the visitors and the housing estates with glowing red pantile roofs, the habitation on Furness across the bay and Piel Castle afloat on its island haven. The immense shipyard hangers at Barrow, the ferry heading out to Ireland and the one coming back in and the one sailboat out on the bay. Last of all, I took away the power stations.

There's a local story that St Patrick, after whom the chapel is named, landed at Heysham bringing the Christian message from Ireland, founding one of the earliest Christian Oratories and communities here on the headland. Vikings moved into Heysham during the 10th century, travelling in from existing communities in Ireland and the Isle of Man. Some of them had been absorbed into the Christian tradition, but some remained pagan. Without doubt, the historical edges here are blurred.

A 1970s inspection of the site found that the chapel was 'rapidly deteriorating' as a result of the severe weather on the headland, vandalism and the constant stream of tourists traipsing through the ruins. As a consequence, human bone fragments could be seen in the earth and the 'underlying stratigraphy was greatly at risk'.[1]

Archaeologists and restoration teams moved in to survey the site and to preserve the ruins from imminent collapse. Amongst the remains they found a Viking woman together with her decorated, carved bone comb. She'd been buried in a shroud beneath a fine spread of charcoal. The presence of the comb infers that she was buried within the Scandinavian pagan culture. Many of the graves were either overlaid with stones or else the bodies had been laid into natural cavities in the bedrock. One of the burials incorporated a stone with a detailed carved bird's head, thought to have originally been a section from a carved stone chair or throne of the style seen in the 7th century. More carved birds were found, both in the headland burials and in the chapel, and on a richly decorated Viking Hogback stone.

This intricately carved stone was found in the chapel ruins and had been exposed to the elements for centuries. In the 1960s it was taken indoors to the neighbouring church of St Peter, built in a dip of the land just beyond the line of trees. The narrative scenes carved on the stone have been variously described as the victory of Christianity over pagan belief, or to show Germanic or Scandinavian legends or myths. On one face, there are wolves, men and a deer; on the other, a tree, birds, saddle horses and a man, possibly with a sword raised above his head. Surrounding all four faces of the stone, a serpent. It may illustrate the Viking legend of Sigurd. In this tale the hero overcomes attack by wolves as told in the poem 'Eiríksmál', written to commemorate the reign of Eric Bloodaxe, the last Viking King of York, and whose rule probably extended over Heysham at about the time the hogback stone was carved. But beware of false idols, say the archaeologists, who argue that a true and detailed interpretation of the stone's carvings is nigh on impossible given that it dates from a time when 'Christian iconography, folklore, convention and mere decoration are inextricably mixed together'.[2]

Me, I like a bit of legend. I think I'll stick with Sigurd.

I like too the symbolism of birds. Looking out from the headland that mid-winter afternoon with the light at its zenith, birds were so much part of the place. Oystercatchers went spinning past, following the edge of the headland, gulls winged languorously out into the middle of the bay, and, on the headland, small birds bounced on pockets of air between the trees. There's a natural justice to their inclusion in the symbolic and celebratory carvings of the time. Given the abundance of birdlife on Morecambe Bay *and* the idea of migration, perhaps the stone-carver thought of birds *and* people, journeying into unseen places, and most, though not all, returning home again.

* *

The first building on the headland was a simple, smaller structure than the ruins that remain might suggest. It had plastered and painted interior walls and a door that faced towards the west and the horizon of the sea. For me this immediately resonated. The Gallarus Oratory in Ireland, 'the church of the place of the foreigners', is a small but perfectly formed and ingeniously corbelled building in the shape of an upturned boat. It sits on the lower slopes of Brandon Mountain, close to the Cosán na Naomh pilgrimage route, its function to offer shelter to those venturing on to the holy summit of Ireland's second highest mountain. I've been to Gallarus many times. Even allowing for all the Germans and Italians doing the tourist rounds of the Dingle peninsula, there's something quietly wonderful about it. If you have the Oratory to yourself, even for just a few minutes, the meditative qualities of the place become clear.

The last time I was there I fell into a conversation with the young Irishman at the gate as he collected our tickets. I'd asked him about the tradition of offering shelter and mentioned the chapel at Heysham. He knew his stuff and was aware of it as another early Christian site.

'There was a whole network of these paths right throughout Europe,' he told me. 'Spreading out from Rome into the east, to Palestine and Jerusalem, through the Byzantine Empire, through Spain and all the way up into Ireland. And there was a great tradition of offering pilgrims food and shelter. What you have here is the end of a major pilgrimage route and so the Oratory would have been a place to shelter, somewhere to find food and to pray on the way up and back again from the summit of Brandon Mountain.'

He talked about the route of the pilgrimage through Ireland, about how routes connected one to another, and how sometimes those undertaking the journey were returning from, or originating from, the far edges of Europe. Their destination this, a mountain on the Atlantic coast of Ireland and from where, if you were to carry on travelling, as St Brendan the Navigator had, you might come to an island in the middle of the ocean that was in reality a giant fish, inhabited by birds and monsters and veiled by mists that would hold you there for years.

The guide's enthusiasm for the subject and his knowledge about Gallarus and its associated history was truly impressive but I couldn't begin to remember it all. I asked him if this stuff was available on a website somewhere.

'No,' he answered, as if taken aback, 'it's all in me head.'

★★

More digging, more discoveries. In 1993 another excavation at Heysham turned up more than 1,200 artefacts, arrow-heads and spear barbs and showed that the area had been occupied from at least the Mesolithic period of 12,000 years ago. To put this into context, the Irish elk was still roaming throughout Eurasia and into Africa and would continue to do so for a further 2,000 years. Land ice was retreating in northern Europe as Earth entered the Holocene period – the warming up of the planet. You might still catch a glimpse of a woolly rhinoceros or a mammoth on your way back from the hunting grounds of Morecambe. That's if you hadn't already speared and dismembered it yourself.

★★

They're found in the turned earth, brought out into the light again by the plough or picked up from a riverbed, mistaken for a cobble. They've been recovered from bogs and springs and streams and trawled up by dredgers from riverbeds. Fifteen of them were found in the River Thames.

Smash one onto the floor and notice the momentary smell of burning as it breaks apart; proof of its origin as the shard of a spent thunderbolt. Keep one in the rafters as protection from lightning; it rarely strikes the same place twice. Put one in the water trough to keep cattle free from disease and increase the milk yield of cows, or use its finely ground powder as a cure-all for illness, to assist in the birth of children or to prevent putrid rot or decay.

Or go for a walk on the beach at Half Moon Bay one day and, having noticed it amongst all the other beach stones, see that the shape of it was just the thing. Pick it up, carry it home and put it to good use as a doorstopper, the thing unidentified and therefore diminished. But then a neighbour called in for a chat, noticed it and recognised it for what it was. They took it away, released it from obscurity and donated it to the museum up in Lancaster. You can see it there now, one of three on permanent display. But hold it. Heft it, feel the weight of the thing and the way it fits into your hand. Look at the colour – jadeite, greenstone – such a dark green it's almost black. Look at the edges, the way it was made, or manufactured. Precision engineering – a measured and balanced gradient from the widest holding point to the fine sliver of the working edge.

Axe: a noun and a verb. Something to work with, something to grind, an action, a transformation.[3]

The very first factory of any kind in Britain, its location at the top of an inaccessible corrie high up in the mountains of the

Langdale valley. A cave hacked out of the rock and measuring six feet wide, seven feet deep and seven feet high, greater than the height of a Neolithic man. The same fine-grained stone was found at other locations too, many much more accessible, but this was the chosen place. From a point 700 metres up on the headwall of Cumbria's Pike o' Stickle the trade in axe-heads began. Imagine then, the mythology that travelled with the axes, and that attached itself to them *because* this was the place of origin.

Francis Pryor writes in *Britain BC* that 'in simple, practical terms, it doesn't make sense', any more than the transporting of the stones of Stonehenge from the mountains of south-west Wales to Wiltshire. But this is exactly what happened. Both locations are difficult to access and no doubt possessed an other-worldly mythology of their own.[4]

Most of the Langdale axes that have been found were rarely used to chop or to cut. Sometimes called 'celts', from the Latin for chisel, they were keystones of Neolithic life. (They have also become, literally, touchstones in the very foundations of archaeology.) Axes from Langdale travelled extensively and the journeys they were taken on and the destinations they arrived at have given archaeologists insights about how and why they were used. A tool, a token, a symbol. The Langdale axes make articulate statements not only about their practical use, but also, and more importantly, about how they existed as part of a system of belief.

> *Axes were statements in stone... They built up biographies and associations and some may even have had names. It was this potential that came into focus when they were singled out for special attention: held aloft or handled as cues in the telling of stories or the recitation of names; treated like those they lived alongside, buried or broken just as bodies were.*[5]

Axes spoke of cleared land, of the very beginnings of farming and were widely used in the creation of order and settlement. Across the 2,000 years of the Neolithic, axes were made and used all over the country from locally available materials. But axes from Langdale are amongst the most travelled. They've been found in Lancashire, Wales, Scotland, in eastern England and at Ballygalley on the north-east coast of Northern Ireland. On a clear day you can see that coast from Heysham.

When Professor Bill Cummins examined nearly 2,000 Neolithic axes from all over England and Wales, a staggering 27% were made from polished Langdale volcanic tuff and found in the greatest concentrations in Peterborough, in Lincolnshire and the east Midlands.[6] In a separate study of axes found in the River Thames, out of 368 specimens a mere 17 were damaged and most were in remarkably good condition. Fifteen of them came from Langdale.[7] Rivers travel through time as well as through the land and may have been regarded as a link between worlds. Axes found in rivers may have been thrown there as part of a ritualistic offering to spirits belonging to a different world and who demanded gifts.

This Heysham Langdale axe, once employed in the mundane task of a doorstopper, was and is, still, a thing of awe and wonder. You can sense the weight of the thing, imagine the way it would fit into a man's hand, or see him hefting it, balancing it in one hand and weighing up the thing's potential. I think he would have spoken with it, about it and valued it intrinsically. The reason it was found on the beach will probably never be known. It may have been lost while on a journey south, east or west. And it's not inconceivable that the people who made and traded axes transported semi-completed pieces to finishing stations further along the route, and perhaps Heysham was part of a chain of trade.

I've been up there to the cave in the headwall of the Langdale Pikes. I've slid and slithered my way down the great stone gully there, the first time as a teenager on a 1970s school fell-walking trip. But I've been back since; it's one of those places that call to you from time to time over the ether, that come in unbidden when least expected. I've walked to the top of both the peaks, of Harrison Stickle and Pike o' Stickle and looked out over the valley that cuts a swathe through time. In fact it was there on one of those summits that I first had any real understanding or sense of time, of how it grinds on imperceptibly, slowly forming valleys and rivers and mountains, and how I, my teenage self, was no more than a speck on the surface of it all. I found it an uncomfortable reconnaissance then. Perhaps less so now. Perhaps.

The school fell-walking club had a different kind of day on offer than the usual walking in the hills and then eating the ubiquitous left-over bar of emergency ration Kendal mint-cake on the bus home. The plan was for ghyll scrambling and scree running. No one thought anything of it then. Having gained both summits we retraced our steps to reach the start of the descent at the coll between two of Cumbria's most characteristic peaks, the Langdale Pikes. Mr Hawksley instructed us in the gentle art of scree running; to descend crab-wise, one leg dropping and the other catching up, like wading downhill fast through snow. Except it wasn't snow, but layers of scree, the jumble of rocks and stones that form where the gradient of the mountain is steeper and where ice and water have made the rock faces crack and splinter, shatter and fall. The headwall of the scree slope had become eroded and difficult, nothing much to hold onto, mainly thanks to hill-walking enthusiasts descending by this route. The surface had become reduced to shale and there was nothing good to grab onto in the event of a stumble.

We found the axe factory. I hadn't known what to expect, but

I think I thought it would be more than this, a small cave dug out of the side of the mountain. But the more we looked, and then found it more demanding than we'd anticipated to scramble up to it, and then measured it against the size of an axe-head, we began to understand. Think of the people who made this. We wore well-designed leather fell-boots; they had skins. Little wonder that Langdale axes acquired status and garnered mythologies that ultimately travelled with them. Stories of arduous climbing, of inaccessibility, of struggle and of reaching a place above the cloud high in the mountains. An other-worldly place that assumed importance above all others. Of the axe makers maintaining their intention to create not only the product but the source and the meaning and the journey. Imagine then, hearing all this in the flat Neolithic fens of Lincolnshire. Of handling a dark, polished, greenstone axe that has travelled along invisible pathways to arrive here in your hands. Turn it over and round and listen to the story of its place of origin. It would be unbelievable.

On the way down the scree we searched for 'rough-outs', roughly-shaped stones that had been tried and rejected. I went down more slowly than the others at first, searching for treasure amongst the tumble of scree. There was a piece I chanced upon. I took it home and kept the thing for years in my room. But it may have been a fancy on my part that once it was held by a man, that he'd chipped and knapped for a while, turning it over and over in his hands to assess its potential before coming to the truth of it, and casting it aside.

* *

'Are you doing a study?'

A man had appeared at my side. I'd heard, rather than seen his arrival, the brakes of his bicycle squeaking as he came to a stop and jumped off. With the slightest nod of his head the man indicated the notebook and pen in my hand. He had a small white dog and, on his command, it took off, scarpering down the hill-side in a way that told me the dog knew the routine.

'Oh no,' I said. 'No, just interested, really. It's such a great place.'

'Oh yes,' he said. 'It is indeed. In fact it's a very ancient and mysterious place. *Very* mysterious.'

As we talked he held the bike by its handlebars and brakes and a continuous high-pitched squeak underscored our conversation. The man's hair was long, grey. He was tall and had the look of someone noble, of someone slightly lost. As we talked he looked out at the bay in front of us and then he gestured, sweeping his whole arm over the headland to drive his point home.

'They reckon this whole headland is covered in ancient burial sites.'

The small dog foraged, following scent trails in the grass.

I saw exactly what was happening. I thought the man probably talked to anyone if he could catch them unprepared. But I didn't mind; the sun was out and I had time. We exchanged a few sentences about the rock-graves, and then, as I thought he might, he hijacked the conversation. It began with the lack of access to library toilets for non-library users, the cost of water rates, the fact that it's impossible to *own* water, or to sell it for that matter, the fact that no one even owns their own name. We were heading downhill fast. I listened, trying to understand how he'd got to this point. But I wanted the day to be positive after weeks and weeks of rain. I wanted the mood to be light. I used what tactics I could muster.

'I know what you mean,' I said, 'but the thing is to know about all that stuff, to think about it and acknowledge it, but then move on. I mean – look at this wonderful place.' And then I swept *my* arm over the bay.

The man paused, lifted his head slightly and regarded me down his long, aquiline nose.

'So,' he said, 'where do you come from?'

'I live in Kendal.'

'Ah. All the way from sunny Kendal. Well, don't get me wrong,' he said, 'you're right but...'

...and on he went. There were peaks and troughs in the monologue but I stuck with it, butting in when I could. Being Cumbrian I'm a self-confessed admirer of banter, in fact I think at best it's a highly developed art, but more, I like people who take on the fabric of oppression and unfairness and who do write to their MPs about stuff that matters.

'I've been writing to the government about how you can't own water. So if you can't own it, you can't sell it, can you? And did you know that when the government write back to you their letters are full of magic words? They're very powerful words too. You have to work hard to decipher it all, to work out the secret codes. *Then* you'll see the *real* message. Once you've done that, you can write back. Play them at their own game.'

It was, I thought, a very mysterious place. I looked out at the bay again and down at the icy graves. The light had begun to decrease in its intensity already. I took a glance at my watch.

'I'd better get going soon,' I said, and detected a slight shift in his eyes, a small sinking of the light. I wondered if once I'd left that would be all the company he'd have today. 'I've to collect my kids from school. I'd better not be late. It's been really interesting talking to you. And thanks for telling me all this.'

As I headed off towards the beach again, the man set off walking down over the Barrows. He waved a long arm in farewell and at the same time made bird-like, fluting whistles in the direction of his dog.

A single oystercatcher piped me back along the shore.

Three

The Morecambe Bay Cockling Disaster

To be honest, it was an effort to drive down to Morecambe on a freezing cold February evening. Although I had thermals on and a windproof coat, after just 10 minutes on the beach I felt as if I'd had enough. Already my fingers were turning numb. I tried to warm them up holding my hands out towards the glowing coals of a brazier, but if I moved even a step or two away, the cold was there again, unconquerable. A group of musicians were playing beside a sculptural boat, its hold packed with a stash of planking and gash wood. How they managed to make their fingers work, or blow enough warm breath into their brass instruments, was beyond me. In canvas shelters steam rose from underneath trays of hot food along with the aroma of Chinese cooking. For such a bitterly cold evening, the number of people marking the 10th anniversary of the Morecambe Bay cockling disaster was impressive.

Away from the main hubbub, a line of ceramic forms snaked along the beach towards the utter dark of the bay. The distance the installation covered must have been 50 metres, and it stopped just short of where the tide washed itself quietly away into the night. The forms seemed to glow, lit only by guttering fire-cans set in the sand at their sides. Sometimes people stopped and crouched down to look more closely at the images and words printed on the surfaces.

I walked the length of the installation myself and squatted down to look more closely. I saw the face of a Chinese girl, the image of snowdrops and Chinese script. Someone had placed a bunch of fresh tulips on the surface of one of them, and printed on another the unmistakeable X-ray image of a truck.

I read the names and statements printed on the side of each one:

'They are threatened by a strong incoming tide'

'She was widowed with a son aged 14 and a daughter aged 9'

'He had a red bag with him containing items for good luck'

'They were identified by good luck charms, watches, wallets, photos and jewellery'

'His father identified a wallet, red bag and yellow cloth with Chinese writing on it'

'Tell my family to pray for me. I am dying'

'In a rich world he was invisible. He became visible only by dying'

With the sand now wet underneath my boots, I reached the final piece. I considered the raw cold that the Chinese cockle pickers would have experienced, the way it gets into your bones and there's nothing you can do to stop it until a warm place is reached again. But of course, there was no warm place for the 24, just the incoming tide, and no one to guide them safely off the sands.

I looked out and wondered what it must have been like, working inside the void of uncompromising, utter dark where the only means of orientation, and the thing that in all probability led to their unnecessary deaths, were the lights of the villages and towns of the coastal communities around the bay. I thought of their hands and the degree of cold they experienced whilst shovelling the surface of the mud. With their workplace lit only by the headlights of their van and a 4x4, I thought of their aching backs after loading pile after pile of cockles into bags. I found it unimaginable.

⋆⋆

As I walked back up the line I met my friend, the ceramic artist Vicky Eden, who had made the installation as her personal response to the tragedy. Together we watched people following the line of sculptures down to the wet sand and back again. Then, with the bonfire boat set alight, we stood in its billowing heat eating Chinese food. A choir sang, silhouetted against the light of the fire and I recognised a reporter from the local TV.

Vicky and I talked about the importance of remembering the Chinese victims, and about why it mattered. It was good to see people looking at that substantial piece of work. When I first heard that Morecambe's community music foundation, More Music, was making plans for an evening of remembrance for the 10th anniversary, I got in touch with their director, Pete Moser, a friend from Ulverston days, and told him of Vicky's installation. The memorial evening was good news to Vicky and me; we'd been having conversations around the idea that there was the distinct lack of an appropriate memorial on the coast, something tangible and significant. We were thinking then of the idea of the possible commissioning of a new piece of sculpture, but then More Music's memorial event was organised and that felt right. It gave the opportunity for Vicky's work to be shown once more. Although she made the installation in 2004, this felt like the right situation for it to be brought out again into the light, or into the dark.

A week or so before the memorial event we met in a café on Morecambe's prom, Pete, Vicky and me and some of the team of artists and performers that Pete had recruited. As we drank our coffee, Vicky told us about the night of the disaster.

'I hadn't been able to get to sleep that night and came downstairs to make a hot drink. I turned on the TV and there was the rescue going on and I heard the helicopters coming, flying low over the house. It was a terrible feeling knowing that someone was in difficulties on the bay, but then much worse when we found out there were significant numbers of people.'

And there's a problem. It seems impossible to verify exactly how many people died that night. Newspaper reports said there had been 23; I heard the number 24 both from other parts of the press and from fishermen involved in bringing in the bodies. I asked the coroner's office. Their reply: 19. The fundraisers, who arranged to clear the families' debts, reported 23 individuals. The chaotic scenes being played out on the bay before and during the disaster appear to have been mirrored by an inconclusive number of dead.

Vicky's installation, 'February 5th 2004', is a series of 23 red earthenware ovoids, each one a footprint of a different place on the bay. The surface of each piece is a cast of the ripples and depressions of the sands from all the places she visited. She added screen-prints to the ceramics on the themes of migration, loss and modern slavery. She asked the Taiwanese artist Chun-Chao Chiu to add the victims' names in Chinese script. She talks about the installation as a pathway from the coastal communities here to the families in China who lost their loved ones. A film has been made about the work and shown at international ceramic conventions. Vicky was brought up in Morecambe and had family who were once involved in the local fishing industry, so her sense of connection runs deep. She recalls family trips to the beach as a child, of jumping on the sands until they turned to jelly, and the ever-present warnings not to go too far out.

★★

I too remembered the news breaking. I knew about the cockling gangs fishing the bay – it was hard not to know about them. On any trip to the Morecambe area when the tide was out, you'd see gangs digging on the sands; cockling had become an industrial operation on an unprecedented scale. The indigenous, small scale and sustainable cockling families were forced to stand back and watch their livelihoods being dug out of existence. Even allowing for all this, I had no idea that workers were being sent out onto the sands in darkness. That was madness.

In the aftermath of the disaster, information and recrimination flew backwards and forwards with the frequency of the tides. Local people – amongst them the fishermen and sand-pilots, who act as guides across the treacherous sands – had been warning of an impending disaster for years. The free-for-all on the sands turned into internecine warfare where gangmasters from the UK were accused of setting fire to bags of cockles collected by the Chinese, and the already disempowered migrant workers were reduced to tears. Adding to that incendiary situation, the Chinese gangs were selling bags of cockles for the rock-bottom price of £5 or £6 a bag. But the demand for shellfish from Spain and France was insatiable; it kept an army of workers going wherever and however they could get at the food.

In the previous year a local fisherman, Harold Benson, rescued a group of around 55 Chinese workers from the incoming tide. In the background the regulators apparently struggled to come up with a workable solution. In December 2003, just eight weeks before the disaster, the cockle permit scheme was introduced that included a safety training course. In the event it seemed far too little, and far too late. In addition, the gangmaster licensing laws were out of date and unwieldy. It doesn't take a huge leap of imagination to guess which year it was amended. That's right, 2004.

Harold Benson was called in by the coastguard to help with the rescue. In the event all he was able to do was help to bring in the bodies when dawn came. It was reported that he was so traumatised by the events that he's never been onto the bay since. He had known though, that there was a safe way off the sands even in the dark. 'Even when the tide hit them, had they had anybody with them, like me, who knows the area, there was still a safe route off the cockle bed… They could have walked to safety.' It seems that, in their panic, the workers had headed straight for the lights of the shore and walked into the deepening channel of the River Keer.

I talked to Alan Sledmore, who led walks across the bay from Hest Bank for many years, and Stephen Clarke, one of a dying breed of fishermen who still fished from nets placed out on the sands, and who also ran the walks with Alan.

'There were hundreds of people out there and they were taking everything. The cockle beds were decimated. I doubt they'll ever fully re-open,' Alan told me.

'In sustainable cockling, only the mature cockles were taken by a few local fishing families and oystercatchers. The only thing that fed on the spat (seed cockles) were the bottom-feeding fish – flukes and plaice. For hundreds of years local fishermen used a "jumbo", a kind of rocking cradle that brought the cockles to the surface. They riddled them and all the immature cockles fell back into the sands. It was the sustainable way to do it. It made it all workable. In comparison, this was a travesty. No one was protecting the cockle beds. Now they're decimated. You used to be able to say, "There's plenty more fish in the sea." But not here, not now.'

I talked to Jack Manning, of Flookburgh, who fished the bay for decades. He told me of the 'terrible cold' he remembered getting into his hands when cockling. He showed me film from the 1950s and '60s, cockling and shrimping with horses and carts, the horses wading into the channel and pulling the load of cart and nets, and the water so deep their noses continually dipped the surface. His son and 17-year-old grandson went to help the rescue. Together they brought in six bodies before it was light. 'That was a terrible thing for a young man to see.'

What had been a very public human disaster also became an environmental one, though not one that reached the wider public consciousness, or that was followed by the media. The indigenous industry that had involved and sustained a few local families had come to an abrupt end.

Stephen Clarke's fishing nets are three-quarters of a mile out, by Priest Skeer, which is a raised area of glacial moraine close to where the Chinese had been working on the night of the 4th of February. After the disaster the local coastguard wanted him to phone in and let them know each time he planned to go out and again when he came back in.

'I just couldn't get in the habit of it,' he said.

I could understand the coastguard's unease, though there's a world of difference between local fishermen with a lifetime's experience and exploited foreign workers who had been instructed to carry on digging after dark, while back onshore their gangmasters waited in the warm for two sets of headlights that never came.

During our meeting, Stephen's phone had rung – his fishing mate. The men had a brief conversation and then Stephen asked a question. 'While you're on, do you know if anyone's cockling on the bay at all these days? No? No, I didn't think so. That's what I thought.'

Stephen's mate confirmed that there are no cockle beds left on Morecambe Bay. Later on I'd phoned the Fisheries to ask about the state of the Morecambe Bay cockle beds. They told me that there are naturally occurring fluctuations in cockle beds, that they hadn't been seen on the Ribble, further down the coast, for 20 years but now there are dense quantities. They told me that the law states that a riddle has to be used, that there *are* enforcement officers checking up, and also that cockle picking cannot be only a local occupation ever again. 'That's just not the way the world works these days.'

In a couple of ironic twists of fate, Stephen had found two unsettling pieces of evidence from the cockling disaster, though there were years between the finds. He told me the story of coming upon what he'd thought was a set of false teeth stuck fast in the sands. He began to dig, and very quickly realised very that he'd found a human skull.

'It was small, and the teeth were very good. Not like our Western teeth. I thought straightaway it must be the skull of the missing woman. I found it on a cross-bay walk. We had to be discreet about how we got it out and back to land. We managed though, and later on it was identified as the missing woman.' Liu Qinying perished along with her husband, Yu Hua Xu. Their son, Zhou, was orphaned at the age of 13. The body of one victim has still never been found.

'Someone said to me that we should have left it there. We should have called out the police or the coastguard. But you can't do that on the bay. By the next day it would have been buried and lost again.

'Most of the bodies were found the next morning, almost all of them on or close to the skeer – that's where they'd been working. If they'd asked one of us local fishermen, we'd have said to look for the other bodies in the channels. They would have been covered over in the space of a couple of tides.'

And then, just weeks before the 10th anniversary of the disaster, a second grim reminder was returned by the sands. A scattering of bright red stringbags filled with cockles had re-surfaced beside the skeer. I remembered the photo and the story in the local newspaper.

'You couldn't make it up, could you?' Stephen said. 'No one wanted to touch them; seemed too much like bad luck.'

It would be mistakenly easy to think we'd learned our lessons. The beds here in Morecambe Bay might be closed, but migrant workers are still digging cockles in extreme places. In 2011, seven years after the Morecambe Bay disaster, a cockling gang were rescued from sandbanks out at sea. They'd been taken out in inflatable dinghies towed by a small fibreglass boat to dig cockles from sandbanks, surrounded by the sea.

★★

Since the cockling tragedy, Pete, the founder of More Music, has been travelling out to Hong Kong and China, forging links with musicians, performers and communities and bringing Chinese artists and musicians to the UK for collaborative work and performances. A community opera evolved, *The Long Walk*, performed in Liverpool, Gateshead, Morecambe and Hong Kong. A couple of weeks after the memorial evening I drove down to talk to Pete about his visits to Fujian province, where most of the Chinese migrant workers originated. Driving along the prom road, the bay was about as full as it could get, and sludgy brown waves sloshed against the sea-walls and breakwater pier.

'The places they came from weren't villages,' he said. 'That's an idea put out by the media. Fuzhou is bigger than Preston; it's an industrial-sized city.'

There were pictures of gaunt, brand new and monolithic apartment blocks in the city of Fuzhou, built to swallow excess profits from China's booming business economy. None of them had ever been lived in – there was simply no demand. A disused holiday resort reminded me of Middleton Sands, once a top destination just down the coast from Morecambe.

Then the image of a vast tidal area, Fuqing Bay, and cockle beds at the edge of the bay with areas divided up and roped off, each site an excavated mud basin with men bent over digging out cockles. The images were astonishing because there were intrinsic similarities to Morecambe Bay. I hadn't anticipated a tidal estuary and cockling industry and it seemed good to see this; at least some of the victims of the Morecambe Bay disaster might have been familiar with the work that took them out onto the bay, the work that ultimately took their lives.

We talked about the aspirations of people who paid, and continue to pay, to travel from China to the West.

'The press gave the idea that these were poor people, those without hope. But that wasn't the case at all,' Pete said. 'To be able to take those kinds of loans out, whether from money-lenders or family members or elsewhere, and to pay a fixer to take you across the world, that was very much a middle-class thing to do; they saw it as a step up. The truly poor couldn't begin to access loans like that or even think of moving up the social ladder.

'When I was there, I was travelling with artists and musicians. They were perplexed about why I would want to raise the issue of the Morecambe Bay tragedy. In Fujian province people either hadn't heard about the tragedy or didn't get why it mattered; in China large-scale industrial accidents happen all the time.

'In many ways, the economy of China is still pre-Industrial Revolution; life is cheap. If 24 people died in the UK, they might say, so what? For me this was a hard thing to hear. We have such a completely different way of looking at the world.'

Pete put the Morecambe Bay disaster into a 21st-century context.

'This was the UK's largest single industrial accident in decades. In China, people hardly batted an eyelid.'

* *

The Morecambe Bay Victims fund was initially set up by the film-maker Nick Broomfield. His film, *Ghosts*, graphically illustrates the life of trafficked migrant workers and, in particular, the story of the Morecambe Bay victims. The freelance writer Hsiao-Hung Pai (whose research was drawn on for *Ghosts*) and the businessman David Tang were also involved in the fundraising efforts that ultimately enabled 22 of the families in China to clear their debts. In 2010 a fundraising walk of 60 miles from Morecambe to Liverpool took place. All the debts were cleared, with the exception of one victim who came from a remote region, and whose family could not be traced.

Three children were orphaned by the disaster. One of the boys went to university in Fujian, and another boy stayed in his home town and found work on construction sites. The orphaned girl went on to study and work in Japan. A further two boys who lost their father were sent to study in Japan, with their remaining family borrowing heavily again to send them, after their original debts were paid off.

⋆⋆

Almost a year after the tenth anniversary memorial event I went to the bay again with Vicky. She wanted us to go early so that the job she had in mind could be carried out discreetly. It was still dark as I arrived at her house. We went on to Bolton-le-Sands and set off walking as the day began finding its colour. The previous day had been torrential – rain all day long, barely ever becoming light. But now the sky was finding another kind of weather, and we set off walking across the jigsaw puzzle of saltmarsh. Forwards and sideways we went to find a way around the deepest, widest pools, then negotiated our way over the jumble of torn edges of the marsh, where great clods of the stuff were breaking down into silt, and finally we stepped out onto the sands.

It was firm underfoot. A meniscus of water remained from the ebbing tide and the light was good. Over Morecambe, vast columns of white towers broke free from a mass of blue-grey cloud-mountains. Around the edges of the bay the landscape and hills receded into pale, ethereal distances; the Furness hills, Humphrey Head, Whitbarrow, Arnside Knott. The big sky was growing lilac-blue and over the bay itself, barely a cloud. Looking down at our feet, the same blue, travelling with us.

Vicky hoped to find a shallow channel. She'd been sinking the clay footprints two at a time. What else to do with them? To me it seemed entirely right: giving back to the bay something of what it had taken, but something also that it had given. We carried a piece each. Mine was a cast of the sands, and embedded in the ripples was a scattering of small cockle shells and the footprints of wading birds foraging. It was a beautiful piece. Beauty and poignancy combined; that all those deaths should have come from the business of picking these tiny, primitive, shelled creatures. I'd written a poem in the aftermath of the tragedy. In it I talked about following a line of bird prints through a shallow channel of water to a place where the bird had clearly taken flight. About how the birds are adapted, comfortable and equipped to be in that environment, and how those others had not been able to take wing, their footprints coming to a similarly abrupt end.

Moss had begun to form in the depressions of the cast and a handful of beech leaves had dried onto its surface. I was carrying, literally, a piece of the bay with me, and it was about to be returned. Vicky hefted a different piece upon which she had screen-printed the front page of *The Guardian* with the words, 'Trapped by the tide and sinking sands in night of growing horror' – one of the earliest reports of the tragedy. Around the ceramic rim ran the headline 'They were Victims of the Sands and the Snakeheads'.

I noticed a low island maybe half a mile offshore; distances were hard to judge. It was a piece of land so shallow that it was barely there. Priest Skeer, close to where the workers had been digging cockles on the night of the disaster. I pointed it out to Vicky.

'I didn't know that's where it is,' she said. 'Well, that's good in a way. It feels like I've been sinking them in the right place.'

'And look,' I said, 'there's the sand bank.' I pointed, tracing a dark line of sand that ran due north from the skeer, and from where the Chinese might just have been able to reach land again.

After a couple of minutes we reached what we thought might be a channel. We'd seen it glowing distantly, a linear stretch of water that followed a curving trajectory towards the south. But arriving there we saw that it was just a slightly deeper skin of water. I wondered if it might have been the line of an old channel. Given the volume of rains we'd been having over the turn of the year, it was entirely possible.

We set the two pieces down in the water. Against the reflected blue of the sky the ceramic forms looked red, earthen, of the land. We looked at them, took photographs and Vicky said, 'I'm pleased there's no sign of the others.'

'Don't worry,' I said, 'they would've gone with just one or two tides.'

We turned to walk back to the shore, the two forms beside each other, the only intrusions in all that flat space. Above our heads curlew and gulls drifted on thermals in the electric blue air.

Four

Things that Were Buried Come Out Again Into the Light

I'm paddling in the sea looking down through clear, shallow water at the sand, and I'm trying to work out why this isn't the real sea, and why it isn't the real sand either. It looks like sand, and it feels the same as any sand I've ever walked on before. Just back there on the grass bank Mum and Dad are sitting in deckchairs, using the car as a windbreak. Dad's wearing a white shirt and tie, formal as ever, even on holiday, and reading the *Daily Mail*. Every now and then he reads something aloud to Mum. She's like Grace Kelly transported to the Lancashire seaside: dark, wing-tipped sunglasses, lipstick, cigarette held aloft between her fingers and a silk headscarf tied under her chin to keep the breeze away.

I'd gone to walk the shore at Hest Bank. When I'd arrived and looked at the setting and the backdrop of the bay, this memory of being in the exact same place came to me. I must have been seven or eight years old. I remembered crossing the railway line at the level crossing and being suddenly there, beside the sea. Once we'd moved to Furness my mother reminded me frequently that the sea of Morecambe Bay wasn't the real sea, and the sand wasn't the real sand either. She'd found our move to the small market town of Ulverston difficult; to her it was a place, as she so frequently put it, at the end of the longest road to nowhere. But Dad had been unemployed for months. He'd worked on the doomed TSR2 fighter and reconnaissance jet at Warton, near Blackpool, and we were living in Cleveleys, just a stone's throw away from what my mother considered to be the real sea.

The TSR2 project was beset by problems and in 1965 the newly-elected Labour government scrapped it, leaving 1,700 workers unemployed overnight. My father was one of them. He spent months looking for work, and was eventually offered a job at Vickers Shipyard in Barrow; a move was inevitable. When Dad began his new job there were months – I don't remember how many – of staying in lodgings and searching for a house. It must have been through the winter as I recall waiting, looking from the window for the sweep of headlights turning into the road. Then the rush of cold as the door finally opened and in the hallway my father charmed small bags of chocolate coins from his coat pockets for my brother and me.

There was a night before we left for Cumbria when Dad took me out into the dark in the back garden. We waited, I can't remember for how long. Dad kept his plan to himself. He must have thought it a moment that any seven-year-old daughter should see. As we waited, he pointed out the Milky Way, the Plough, Orion and showed me how to find the North Star.

Then something came gliding across the northern dark, but it didn't fizzle out like a shooting star.

'There it is! Can you see it?' Dad said, his arm reaching skywards, following the object through space.

And I did. Telstar, the first transatlantic satellite moving on a pre-ordained trajectory and passing above our house like a prophecy that life would never be the same again.

It seems, then, that I have Harold Wilson to thank for my connection with Morecambe Bay and with wild places. But the place where I found my mountain feet was a place my mother was never able to settle. She pined for our house near, what was to her, the real sea and the beach made of real sand. She missed the simple, immutable coming in and going out of the tide, the real waves that my brother and I could splash about in wearing inflated plastic swimming rings around our middles, pretending we knew how to swim. She wanted nothing more than to take the tram and rattle and spark along the coast to Blackpool Central and the shops. She longed for the prom and the rock shops with striped awnings that flapped and snapped in the sea wind, the rusty red tower, the fish and chips, the buckets and spades, and the donkeys. She loved the salty, seaside gaudiness of it all.

The pictures and memories took a moment to clear, falling away more slowly than they had arrived.

★★

It was April, and one of the first warm days of the year. I bought coffee at the beach café and sat outside where people leaned back in their seats enjoying the air and the view with tea served in china cups. Under the tables next to me two Labradors lay flat to the ground, one brown, one black. They eyed each other up, nose six inches from nose. There was a notice in the café window for a dog lost a year ago over the bay at Bardsea, last seen running after a gull and heading across the sands towards Flookburgh. I wondered if he was ever found, or ever caught his gull.

Ulverston was just there across the bay. It seemed so close in the soft spring light, its houses rising up the hillsides and the lighthouse that isn't a lighthouse on top of Hoad Hill. Recently restored, it stood out from the unfurling hills and backdrop of mountains in its new white paint.

On the empty bay cumulus shadows of charcoal grey migrated slowly across the sands. Where Morecambe Bay and the Irish Sea met, a white sail effervesced through haze and three small spheres of cloud rode the pale turquoise sky. Above my head the colour intensified into the promise of summer cobalt. At the bay's most westerly point the castle on Piel Island rose through the shimmer line as though drawn by hand, a series of short pencil marks that sputtered out to a line of scribble.

The warning siren for the automated crossing sounded and when the gates clanged shut a few seconds later a train blasted past heading north. Another memory: my brother and me up on the footbridge, holding onto the wrought iron railings as a train rumbled underneath, becoming momentarily lost to each other in clouds of steam that surprised me; I'd imagined they would be soft, billowy and cool, not acrid enough to take your breath away.

I finished my coffee and walked up onto the bridge. Take your kids up there now and they won't even be able to see the trains. It's been boxed in with sheets of corrugated green metal higher than a tall adult's head.

I walked onto the foreshore and found deep deposits of cockle shells, the ground a silty mix of broken shells and mud, saltmarsh and pools of water left by the tide. For a moment I was surprised. I was so used to these liminal places being the usual mix of estuarine mud, sand and saltmarsh. But a moment later I made the connection. Hest Bank and Red Bank, just further up the coast, were both places where the unregulated gangs had come to raid the 'Black Gold' of Morecambe Bay's cockle beds. A fisherman told me later that there had been burger vans, mobile loos, refrigerated lorries, a whole unofficial infrastructure to support the new, non-local cockling economy. After the beds had finally been decimated years after the cockling disaster of 2004, the profiteers vanished like ghosts in the night, leaving behind a mess that cost the council tens of thousands of pounds to clear away.

The shadow of a gull glimmered across the sea-pools in the saltmarsh, and transparent shrimp darted for cover from mine.

A narrow channel of water wound from the shore into the bay, moving onwards in lazy curves. A distance out, there was a man walking two dogs on the sands, one black and one that should have been white, but had four black legs and a tideline of black up to its underbelly. They were walking around what looked like the remains of a groin system. It was a structure built of huge, dressed sandstone blocks and with broken stumps of sea-blackened timbers rising from it. I found an interpretation board close by. The structure was the re-exposed remains of Hest Bank Wharf, uncovered by a series of high tides in 2001. At certain places retreating glaciers dumped their loads of boulders on the bed of the bay, creating raised areas known as skeers. The wharf builders took advantage of this glacial load carrying, building an offshore harbour close to the shore. This gave a point of access to shipping that was near to the Lancaster Canal, a few minutes uphill by horse and cart.

The wharf had been completely buried by the sands, lost to generations before becoming re-exposed again. I liked this idea, of how the tide creates surprises, taking things away and then giving them back at some future moment, re-inventing, or re-interpreting the past for us.

But of course these finds thrown up by the sea are not always glamorous or archaeologically important. Walking along a wild, four-mile Atlantic strand in southern Ireland recently, my son and his pal came upon something buried in the sand, and called me over. We peered down at a black, partly-revealed electronic gizmo, and began digging away with our hands to get at whatever it was. As we scooped, we began to reveal white fabric. Convinced that we'd chanced upon a high-tech weather balloon, or a piece of scientific hardware, we dug away. A few minutes later, as more of the fabric was revealed, my son's pal sat back on his heels and announced his verdict: 'It's an air-bag.' And so, we admitted reluctantly, it was. Our expectation of the romantic, the curious, the exotic, was reduced to the mundane. Someone had trashed a car on the beach and here, buried and revealed again, was one of its scattered, constituent parts.

★★

I love the idea of fully-rigged sailing ships coming up into the bay. The way the channels and the whole floor of the bay have silted up though, the most you see these days are small weekender sailing boats, and not that many of them.

In the way of these things, I'd found a piece of evidence of shipping in the bay in the most unlikely of places. I'd been to visit my father in his nursing home in Lancashire. He'd had to move bedrooms closer to where staff could keep a better eye; he'd been falling more frequently. During the move, it seemed the staff had thrown Dad's possessions back into the drawers, so that personal papers, belts, pyjamas, newspaper cuttings, hearing aid batteries, Christmas cards and more had been stuffed together. I was sifting through, making sense and order, putting things back properly and making piles to keep and piles to chuck out.

Then there it was: a black and white photograph of a sailing ship far up into the northernmost reaches of the bay, berthed in Ulverston Canal, a two-masted schooner with her sails furled and a lighter held fast by ropes at her side. Her prow pointed eloquently towards the bay, speaking of the next voyage out. In the background, and at the end of the canal was Hoad Hill and the lighthouse without a light upon its summit, both significant places for much of my childhood. The photograph came to light – unearthed, if you like – in much the same way as Hest Bank Wharf, after being buried for years. But my dad wasn't known for his interest in local history.

'How did you find this, Dad?'

I handed him the photograph. He took it with his curled-over fingers and after a moment or two of peering at it, he'd registered the image.

There was a 'ha' of recognition, and then, 'One of my pals from the U3A found it. He knew I'd lived in Ulverston; thought I'd like it as a memento.'

'It's wonderful, Dad, a really great bit of history.' He nodded and offered it back to me. I turned it over.

The photograph was very old, and had been attached to someone's photograph album. Instead of the name of the ship or of the photographer, there was just ancient yellowed glue smeared and dried on the back.

'Can I keep it?' I asked, and he told me, 'Of course you can.'

Back home again I began to search for information about the ships that sailed the bay. Almost immediately I found the same photograph on a website, and the ship's name too. She was the *Annie McLester*, launched at Ulverston's Canal Head shipyard in 1866, and part of a fleet of ships known as the Duddon fleet until her loss at sea in 1891 after striking Big Scare Rocks off Luce Bay in Wigtownshire. She broke apart and the bodies of two men were found, washed ashore down the coast at Port William. Of the other two crew members, there was never any trace.

I found a story about her too, of how another schooner, *The Ulverston*, berthed alongside the *Annie* in the Ulverston Canal was being loaded with gunpowder. She had caught alight with 80 tons of powder aboard. Two of the crew members from the *Annie McLester* saw the fire and boarded the ship to put the fire out before it had a chance to spread to the hold. I wonder if they were ever rewarded for this, or were just mightily relieved.

I keep the photograph of the *Annie McLester* in my writing room, wrapped in paper to protect it, but writing this I brought it out again into the light. There's a nice triangle of belonging associated with it, this piece of buried treasure. It has my dad's hand on it, and it shows the place we lived and where I grew up, where I had hills and rock slabs and a lighthouse that had no light for a playground. And then it became a subject for me to write about.

**

Following the shore northwards from Hest Bank it wasn't long before I came to the sign that marks the starting point of the historic nine-mile cross-sands route to Kents Bank, the traditional way into and out of Furness before the railways came. The message on the sign was unequivocal:

The Public Right Of Way Across The Bay To Kents Bank Crosses Dangerous Sands. Do Not Attempt To Cross Without The Official Guide.

Before the railway arrived in 1857, linking remote Furness at last to the rest of the country, crossing the sands was the major route to the north. In the Middle Ages the monks who controlled the economic strongholds of Furness Abbey and Cartmel Priory instigated the necessity of guides on the sands. The monks' whole way of life, their trade and their communication with the outside world depended upon crossing the sands into and out of Furness. The first guide, Thomas Hodgson, was appointed in 1548, and since then the position has been a royal appointment. Cedric Robinson is the 25th person in the job, and has held the title 'Queen's Guide to the Sands' since 1963. Cedric and his wife, Olive, live in Guides Cottage at Kents Bank, a traditional Lakeland farmhouse next to the shore and the railway that skirts the bay. The house, four fields and the handsome sum of £15 a year are his payment for guiding people safely across the sands, for keeping the old way open.

There are, though, other guides, each one having specialist knowledge of the particular area they serve. Until recently Alan Sledmore and fisherman Stephen Clarke led walks from Morecambe or Hest Bank over the long crossing to Kents Bank. The guide for Ulverston Sands, Raymond Porter, fishes off Canal Head and holds walks from Conishead to Chapel Island, and John Murphy from Barrow leads walks from Walney Island across to Piel Island.

★★

I came to a small headland with low, exposed sandstone cliffs. Wind-sculpted thorn trees gripped the steep slope, arching their dense branches towards the gradient, lit by the merest suggestion of new green. This odd, late spring had us all talking; the nonsense of snowdrops out at the same time as daffodils, and even bluebells popping up simultaneously in small numbers.

A small group of walkers were following the footpath that traversed the crest of the headland. They were walking away from the sea, towards the hills. The landscape at the edge of the bay speaks of this slow shift of environment, each segueing into the next. It evolves from rocky shore to farmland, from saltmarsh to sand and from limestone outcrops at the edge of the sea to the bluffs of Arnside Knott and Warton Crag. Further north, at the head of the bay is the cliff-edged escarpment of Whitbarrow. Looking out from the shore there was a strong sense of the arms of the land growing and enfolding the whole bay.

Bumblebees, awake at last, surfed the air, passing me like motorcyclists making lazy swerves round bend after bend. I noticed bladderwrack, a sign of the real sea. Inland, a train sounded its horn, and as I sat on the bank to eat my sandwiches a skylark began to splash its song across the sky. The heady scent of gorse came drifting down from the top of the headland. The sounds and sensations of late spring and a nod towards the summer were beginning to appear.

There was a small wood close by and birdsong was spilling out; it was all going on in there. In the warmth of that morning birds were eyeing up the opposition, picking up signals, proclaiming themselves taken or available. And let's face it, bird song is really all about sex. It's about courtship, mating and territory and the primal urge to procreate, to bring about new life, or to put it another way, to protect and survive. Above it all, a blackbird sang to the morning, as though asking questions, and answering them himself. He seemed an optimistic, glass-half-full sort of a bird.

I began to retrace my steps back along the shore, and then I sensed a slight movement amongst the stones in front of me and I stopped, dead still. A wheatear was picking about in the gaps between stones, apparently unconcerned by how close I was to him. I watched as he worked a radius over the pebbles and through the tideline. The circle he made as he hopped was perfect, as if drawn with a school compass. His tail bobbed constantly as he investigated the ground.

It's likely that the name wheatear originates from the Norse for 'white-arse'; there's a telltale splash of white across the rump. They also have a dark, almost exotic eye-stripe. This one, a male with a bright blue back, was completely indifferent to me, and continued to investigate underneath a large boulder in the middle of his route.

The wheatear makes one of the longest migration journeys of any small songbird. They over-winter in sub-Saharan Africa and journey north into Europe, Greenland, Asia, Canada and Alaska to breed, then return south a few months later. This bird might've only just arrived and was in need of a good meal. Or perhaps this was a stopping-off point, like a motorway re-fuelling stop for a much longer journey deeper into the northern summer.

Over the bay the sky was empty and blue. The cloudscape had moved outwards, so that cumulus cloud streets formed over the land. An older couple walked the shore in front of me; the woman wore pink and the man blue. They walked hand-in-hand, laughing as their small, brown and white patched terrier jumped the sea-pools, skittering about and chasing sticks at the speed of a small rocket.

A moment came when I felt a shift of emphasis in the air, and looking into the distance towards the sea there was a line of movement, travelling at speed. The leading edge of the tide, or the bore, was coming in as a spreading strip of white with the sunlight catching its prow. The line shifted, forming, breaking, shifting and forming again. Raucous gulls began to move out from the land into the bay as if announcing the sea's arrival.

The sea was rounding Morecambe pier and sliding in towards me. Behind the leading wave the water was the sienna colour of unfired clay. Oystercatchers came winging in. The sense of movement was palpable, the water moving, pulsing, a shimmer here, another over there.

At the water's edge black-headed gulls rode low thermals. They were slow-winged, nonchalant. They banked and turned, landed, lifted and moved on, all their movements mirrored by oystercatchers. The temperature dropped again as the wind increased, pushing the sea in front of it.

A pair of gulls took it in turns to rise and fall, one over the other, tumbling and ascending again into the blue. Out of nowhere a cloud of dunlin appeared, *peep-peep-peep*ing their insistent call. Moving as a cohesive whole, they broke apart then coalesced again. Another group came in, low to the horizon. They were an arrow arcing, shooting, measuring the distance. In 30 seconds flat they'd quartered the whole bay, passing close in front of me and doubling back again, crossing the sky where seconds ago their wing-beats broke the air. They came in like distant messengers from the sea.

As the water came, slivers of sandbanks were left uncovered and larger areas too. Where the water filled, the colour changed to silver-grey. As the tide progressed towards Arnside, the water began to slap against the boulders out on Priest Skeer, another glacial slag-heap further out from the shore. The water continued to bounce off the boulders and from the shore it was as if white flags were being raised and lowered, raised and lowered.

Another smaller group of waders came speeding in close formation, then 50 or more oystercatchers materialised with black-headed gulls as outriders. They moved with such speed, like specks of paint flung from the end of a painter's brush. The air was suddenly filled with birds, their sounds and their patterns, as if they were taking part in some wild celebration for the return of the tide. Rising and falling in wave form, they moved as fish-shoals move, turning in the ocean; that singular, forceful choreography.

The sea had pushed beyond Arnside Knott on its way to the northern reaches of the bay. Two fingers of sandbank remained though, waiting to be engulfed. Abruptly, all the birds seemed to have dematerialised and then the sea had arrived in its entirety, no scraps of sand remaining. The blackened remains of the wharf had disappeared again underneath the wind-powered water. I walked up onto the grassy bank, the place my parents had sat in their deckchairs in the 1960s.

I turned for one last, lingering look at the bay, and saw that the birds had landed again. There were so many of them, out at the sea's edge, and in the intense sunlight reflecting on the edge of water it was as if the sea was fizzing. The siren at the level crossing sounded again, announcing another high-speed train. As always, it was hard to leave the sea. The real sea, for what else could it be?

Five

David Cox – The Road Across the Sands

I'M LOOKING INTO A PAINTING. In it there's a straggling line of people, horses, carriages and carts travelling across the sands of Morecambe Bay. It's as if they're travelling to the very edge of the world. On an indistinct horizon, where the sands and the sky merge, an ominous squall builds strength and carries with it fast-moving grey cloud. A mere ghost of the sun glimmers through air made opaque by blown sand.

In the middle ground is a passenger coach pulled by a pair of horses. Four wagons, loaded to toppling height, follow one behind the next. There's a sense that at any moment one of them might lurch sideways, its wheels bedding into soft ground and becoming fast in the sands.

A horse with its head held low labours underneath the weight of two riders. A woman walks holding a basket on the top of her head and she struggles to keep it there; she's frustrated, at her wits' end. Tucked in beside her a young girl attempts but fails to shelter from the weather against the older woman's body. Their clothing ripples like fast-moving currents in water, constantly distorted by the head-on wind. Both are barefoot.

The more I look into the painting, the more I become aware

of the gathering strength of the wind that moves over the surface of the bay. I can imagine the cut and the bite of it; the way that bodies cool down rapidly in exposed places.

'Market Traders Crossing the Sands' is one of dozens of paintings and drawings made of Morecambe Bay by the 19th-century painter David Cox (1783–1859), a major figure in the 'Golden Age' of British watercolour painting. From the end of the 1700s until the early 1900s, British artists working in this medium were in demand across Europe. Watercolour had become cutting edge. In France artists like Delacroix were looking north to Britain and emulating the light, the sparkle and luminosity that painters here were achieving through working in layers of transparent wash. Cox and his contemporaries J.W.M. Turner (who also travelled north and painted the bay), John Sell Cotman and later the Pre-Raphaelites John Ruskin and Whistler opened up possibilities in paint that were entirely new, different in every way from the properties of oil paint. And watercolour had major advantages over oils: it was lightweight, portable and quick-drying and was therefore wholly practical for artists who wanted to travel and make work away from their studio. The new breed of pleasure tourists sought out small-scale images on paper as ideal mementoes of their travels – the postcard or photographic equivalent of the time.

Watercolour was also the perfect medium to show the mutability of our weather: 'its fleeting and shifting atmospheric effects, subtle transitions of light and half-light, and the opalescence of water, clouds and mists.'[8] This description by the art historian and watercolourist Dr Patricia Crown could have been written with the paintings of David Cox specifically in mind. In a 2009 review of a David Cox exhibition, the *Telegraph* critic Richard Dorment described Cox as 'a painter of sudden showers, big skies, scudding clouds and sun rising through mist over wet sand'. He argued that Cox's depictions of people fighting their way through 'empty landscapes under driving rain' were a realistic representation of the hard nature of travel at the time.[9]

Numerous times Cox painted the bay, its weather and its people. From 1834 he spent six years visiting, journeying out onto the sands and recording what he saw. He documented the lives of those who travelled across the bay or who worked on its surface. These paintings are truly elemental.

Another painting. A day of dense air with a sun that only just breaks through, glimmering distantly. Two horses, one ridden by a man and woman and the second, a grey ridden by a man who has an air of knowledge about him. It's probable that he is the guide-across-the-sands. He points to a group of ghost travellers who materialise from the opacity of mist. His dog, ever alert and with its tail held high, appears slightly unsure of what it is exactly that is coming into focus in front of him. A carriage, possibly two, and a couple of people walking and carrying loads are coming towards the guide. They might be emerging from the waters of the River Kent. A flock of gulls land and lift, unconcerned by the weather.

Search for David Cox paintings and they ping up in different places. One of the 'Lancaster Sands' oil paintings sold at Bonham's in London in 2006 for £20,400. Again, a group of travellers move towards the front of the image. Some have covered themselves with blankets. It's raining, and it's cold, wet Cumbrian rain. The light is flat and grey. The tide follows the travellers, almost snapping at their heels.

Cox's paintings hold a lens up to the social history of the bay. But I found another piece. 'Lancaster Sands, Morecambe Bay' shows a group of cockle pickers being called in by their foreman, who is blowing his horn as bad weather approaches and the tide turns. A group of exhausted-looking, silhouetted figures wade through an area where the sea has already reclaimed the bay. It's impossible to ignore the resonance that travels on from 2004.

**

Cox lived through an age of change on an unprecedented scale. The Industrial Revolution had bedded in, people had moved into towns as work patterns changed, and yet for many living in the Morecambe Bay area subsistence farming and fishing remained the basic economy. Transportation of goods was essential and the cross-sands route was the lifeline for many. It was, to put it simply, the M6 of its day. Smallholders from Furness depended on the route to travel to the economically important markets at Lancaster, and for people travelling on to Scotland there was no alternative but to come this way. The much longer journey by the edge-lands was considered even more dangerous; it crossed rivers and bogs and exposed travellers to potential robbery.

William Wordsworth regularly accessed the Lakes by the cross-sands route. I like this, the idea that he set foot on dry land again on the shore adjacent to my home town of Ulverston. He described in almost blissful tones a journey he made in perfect weather, chronicled in his long autobiographical poem, *The Prelude* (1850):

Over the smooth sands
Of Leven's ample estuary lay
My journey, and beneath a genial sun,
With distant prospect among gleams of sky
And clouds and intermingling mountain tops,
In one inseparable glory clad...

**

There's an ink and watercolour drawing, a sketch that Cox made from direct observation, called 'Crossing Lancaster Sands'. The drawing shows a group of traders and travellers leaving the shore.

They move onto the sands when the sea is only just retreating. The first people are walking or riding through the remnants of the tide. Setting off like this and at this point in the sequence of the tides gave the longest possible number of hours for the crossing, a tight six-hour period between high tides. Even so, the very idea of their journey seems biblical to me. With the inherent dangers of the tides and the state of the rivers, each journey was, in its own way, a voyage into the unknown.

Cox made numerous paintings with the title 'Crossing Lancaster Sands'. He reworked the scene again and again but always with this raggle-taggle line of people crossing from one side or the other. He shows the workers, the weight of loads carried and the general sense of struggle involved. In this series of paintings and drawings those travelling in opposing directions meet one another, talk and bear witness to the weather coming into the bay from the west. They might be turning and looking back over their shoulders to where sinister clouds encroach and will in a matter of minutes expunge all the blue. The clouds are loaded with heavy rain and push strong winds in front of them. There's a phenomenon in the bay area of the weather changing with the return of the tide. It may clear out to unbroken and ethereal blue, or bring cloud and rain. A south-westerly wind pushes the tide harder and faster. So the cross-sands guide, frequently shown in the paintings, will be delivering an unequivocal message: the travellers need to hurry, the tide is on the turn, the weather is deteriorating. Time is running out.

The guides had expert knowledge of the sands and the state of the rivers that were an unavoidable part of the journey across the bay. Working on horseback, they waited out at the river's edge to ensure that all travellers, whether on horse or on foot or in carriages and coaches, could cross the rivers safely. There was a guide for the long crossing from Hest Bank to Kents Bank, and another for the shorter crossing from Flookburgh across the Leven to Ulverston. They then pointed out the safe direction of onward travel, avoiding quicksand and gullies over the remaining miles to the shore.

In Morecambe Bay there is great beauty at work but there's also continuous inherent danger, and this links to the idea of the northern sublime. British writers taking the Grand Tour in the 17th and 18th centuries began to write about the notion that landscape can be appealing and beautiful whilst being simultaneously dangerous and something to be held in awe. One of the first to write on this subject, the Englishman John Dennis, wrote about his journey crossing the Alps in 1688 as being a pleasure to the eyes but also 'a delightful Horrour, a terrible Joy'.[10]

Much of the public's awareness, or consciousness, of Morecambe Bay over the centuries comes from this same place, the idea of extreme beauty and terror in combination. The traveller or fisherman working the bay must have knowledge of tides and the changing nature of the sands, and most will have grown up within this tradition. Of tales of deaths, drowning and tragedies there is no shortage – a simple search brings up story after story, and in the churchyard of Cartmel Priory there are tombstones to travellers who perished during the crossing. The writer Elizabeth Gaskell (1810–1865), who lived for a time and holidayed beside the bay at Silverdale, wrote in her short story *The Sexton's Hero* (1847) about the notion of true heroism.

In the story the travellers are late setting out due to the clocks being wrong, and they needed to cross and cross back again during the same six-hour low-tide so that the character Letty can get home to feed her young baby. 'The fresh' has come down (recent rainfall on the mountains filling the river that will have to be forded).

From Bolton side, where we started from, it is better than six mile to Cart Lane, and two channels to cross, let alone holes and quick-sands. At the second channel from us the guide waits, all during crossing time from sunrise to sunset; but for the three hours on each side high–water he's not there... He stays after sunset if he's forespoken, not else. So now you know where we were that awful night. For we'd crossed the first channel about two mile, and it were growing darker and darker above and around us, all but one red line of light above the hills, when we came to a hollow (for all the sands look so flat, there's many a hollow in them where you lose all sight of the shore). We were longer than we should ha' been in crossing the hollow, the sand was so quick; and when we came up again, there, again the blackness, was the white line of the rushing tide coming up the bay...

By this time the mare was all in a lather, and trembling and panting, as if in mortal fright; for though we were on the last bank afore the second channel, the water was gathering up her legs; and she so tired out! When we came close to the channel she stood still, and not all my flogging could get her to stir; she fairly groaned aloud, and shook in a terrible quaking way. Till now Letty had not spoken; only held my coat tightly. I heard her say something, and bent down my head.

'I think, John – I think – I shall never see baby again!'

Inevitably Cox's scenes of people walking the sands bring to mind images of today's cross-bay walks. With people raising money for charities, the walks are hugely popular with anything up to a couple of hundred people or sometimes double that on a good day. Cedric Robinson, the current Queen's Guide to the Sands, leads his people on with a staff in his hand like a latter-day saint. People who take part in the walks will tell stories about it afterwards. About the scale of the bay when you're out there, all that empty, unfamiliar space, and crossing the river maybe up to waist-height depending on how much rain had fallen the night before.

★★

Cox's ascendency to the world of high art came from unlikely beginnings. He was born in Birmingham to a blacksmith father and a mother who, though a miller's daughter, was a forceful and educated woman. Instead of the expected trajectory of a life following his father's footsteps in metalworking, Cox's interest in art was encouraged. There's a story that as a boy recovering from a broken leg he busied himself by painting paper kites. (It seems then that the sky had always held a fascination for him.) At the age of 15 he was apprenticed to a painter of miniatures producing portraits for the merchant classes. He went on to the position of assistant at the Birmingham Theatre, grinding pigments and preparing canvas ready for scene painting. By 1802, after just two years as an assistant, he was given the role of lead scene painter with his own team of assistants and was credited in the theatre's programmes.

Later, his paintings became pieces of drama in their own right. He lived in London and Hereford, wrote books on painting and drawing, and was in demand as a teacher. I wonder about his parents though, what they made of their boy's meteoric rise into the hub of society.

Cox travelled extensively in Britain, including through Wales, Derbyshire, Herefordshire, Lancashire and the Lakes, where he gathered information, drew and painted. His pastoral landscapes in oils bring to mind Poussin, Claude Lorrain and Constable, even Leonardo da Vinci. In these pieces of work, small figures are spotlit but insignificant against the epic scale of the landscape they move through. There might be ruins. There'll be woods and trees alive with weather, and forests, mountains and lakes. People will be at work on the land, and it was hard work, unsentimentally shown.

In a number of these paintings there's a sense of paint applied at speed, of the quick change of a brush and the need to get it all down before the light changed, as if, in contrast to the high art of his more formal landscapes in oils, Cox allowed himself the freedom to paint things that *really* interested him and that animated him. These are places that fed him and gave him energy, and there's a sense of the style moving towards impressionism. The works are simpler, detail has been stripped away and there's a frisson of excitement about them. As with the Morecambe Bay paintings, there's a sense of tension about what might happen next.

In 'Cottage on a Hillside' grey-white clouds are moving at speed, disintegrating and revealing a high cold blue above. Just a handful of brush-strokes and the clouds are complete. A woman hangs a line of sheets and garments out to dry on the line beside her cottage and over the hedge bordering her cottage garden. She is dwarfed by the washing, almost subsumed by it. There is so much of it; the washing on the line snaps in strong gusts of wind. The wind lifts the washing up and the woman struggles to control it. She wants it back in its place on the hedge-top, but the wind has other ideas. This was a woman whose business was washing, and she needs to take care of it. At any moment the piece she's struggling with might be snatched from her hands and borne away. A chemise might lift at the edges before being worked free despite the woman's best efforts. The body and sleeves could become lifted and filled with air before being carried away, a shirt, worn by the wind.

In the painting 'Night Train', made in 1849, the coming of the steam train collides against a more innocent world. In a sky almost as turbulent as a scene from a John Martin landscape, a quarter-crescent moon breaks through ultramarine clouds. A steam train intrudes into this pastoral landscape along a horizon blurred by low cloud. A white horse careers away from it whilst another, more solid horse watches alert, impassive. There's a sense of the edge of land segueing towards a flat expanse that is entirely like Morecambe Bay. The engine's boiler is hot and red, an ember distantly glowing. Steam flowing over the train's carriages has been made by the technique of 'scratching out', where the artist has used a sharp point to scratch the paint carefully away to reveal the white paper underneath. 'Here,' Cox might be saying, 'soon enough these trains will put an end to the very idea of the cross-sands route. It will become scratched out, an anachronism, obliterated by progress. This, right here, right now, is the future.'

Six

Sea Charts and Life Maps

I WANTED TO LOOK at the sea charts of Morecambe Bay and see for myself how the sea bed has changed over the centuries. I arranged to meet one of the curators at Lancaster's Maritime Museum. The museum is housed in the city's former Customs House, a grand affair designed by the architect and cabinet-maker Richard Gillow. It's all imposing Doric columns, porticos and a double flight of stone steps. In its former life it served the 18th-century shipping trade, and as I walked towards it, the only person on the quayside, I imagined the scene 200 years ago, a different kind of skyline made up of timber and sailcloth, of bowsprits and masts, yards and deck rails, all action, sounds and the smell of pitch.

★★

The earliest record I'd previously found of Morecambe Bay was a map of Lancashire made by the cartographer John Speed in 1610. Although it didn't focus solely on the bay, it did show three major areas of sandbanks and two channels that were clearly navigable to shipping: the Ulverston and Kent channels. During my phone conservation with the curator ahead of my visit, she'd told me that a sea chart of the bay dating from the 1600s was her favourite item in the museum's collection.

Michelle came to greet me and took me behind the scenes through alarmed and key-coded doors, then she disappeared into a store-room full of shelves and racks of paintings. She came out again holding two framed charts and placed them side by side on a large table. The earliest was dated 1689; I was relieved that it was behind glass. The second chart was dated 1783. We spent a few minutes looking at them together, comparing and pointing out changes like a child's 'spot the difference' cartoon.

A compass at the centre of the earlier chart sent out radiating lines of directional bearings across the map. North was a small arrow that pointed towards Scotland. The Isle of Man tilted diagonally above the compass, and the coasts of Wales, England and Scotland were arranged precisely around the sides and base as if three sides of a wiggly rectangle. Dumfries and Galloway were reduced to a bumpy line like the outline of a treasure map. Morecambe Bay was placed centrally, just north of the estuaries of the River Mersey and the Ribble, and was the most significant feature on the north-west coast of England. Three engraved galleon-like ships, with their sails a-billow, voyaged across the Irish Sea. They listed to starboard in a strong north-easterly wind, everything made of paper and ink.

Four distinct areas of sandbanks were marked on the chart: Lancaster, Kent, Middle, and Cow Sands. But the most noticeable feature was a navigable channel that wound its way up into the far north-western head of the bay to the small settlement of Green Odd, now Greenodd. It was confirmation, were it needed, that ships regularly sailed up into the furthest reaches of the bay. Down at the north-western edge, the town of Barrow had yet to appear. It didn't emerge until the iron trade began to develop around 1839. But Lancaster was marked, though still isolated from the sea until its development as a port in 1750.

Both charts brought to my mind the illustrations and maps

that Pauline Baynes made for C.S. Lewis's *The Chronicles of Narnia*, images that held me in thrall for much of my childhood. In both charts the ships that sailed across the paper seas appeared to be caravels, the same smaller, more navigable ships that the Portuguese had also developed to facilitate exploration into the African interior.

The larger chart of 1783 was engraved with artistic flourishes and embellishments that would have been commonplace in its day. Cherubs and gods blew winds across the bay and the contours and outlines of hill ranges were shown as sight-lines, aids to navigation when sailing up the channels towards the north. Michelle handed me a magnifying glass, and through it a world of tiny anchors opened up. The map-maker, Samuel Fearnon, indicated 'the place proper for Vessels to lie afloat or a Ground'. Depths were marked in fathoms 'at Low Water in Spring Tides' – scores of them indicated the routine world of seafarers navigating up the channels towards Ulverston, Greenodd and Milnthorpe.

Although the shipping channels were familiar to sailors, Fearnon acknowledged the inherent dangers of the bay. He wrote, 'with proper Directions to shun all dangers and sail into any Harbour, Road etc, contained herein... Humbly described by his most obedient servant...' Further out towards the Irish Sea, depth markers were shown in the place where the white blades of a windfarm now turn. The make-up of the sea bed was described like a user-guide for enthusiasts: 'Stony Ground, Low Water Ebbs', 'Brown Sand and Shells', 'Fine Sand', 'Brown Sand and Red Shells', 'Course Sand and Black Specks', and further out the more ominous 'Soft Black Ground' or 'Soft Mud'.

The bay's infamous sandbanks were mapped with much greater detail. An engraved sailing ship sailed through the Furness and South Deeps towards Ulverston and a navigable channel wound its way up the bay to Grange as well. Greenodd had become a significant creek-port for Lancaster, a place from where valuable minerals were carried out to the wider world – copper ore dug out of the Coniston hills, locally quarried limestone, and gunpowder from the nearby settlement of Backbarrow. Incoming goods recorded in historical documents of the time were raw cotton, coal and sugar.

If you go to Greenodd these days there's a busy dual carriageway that passes alongside the bay for a mile or two. You can mark the state of the tide by driving along, but you'd have little sense that in Fearnon's day Greenodd was a major shipbuilding centre, a place where vessels of up to 200 tons were built. In 1783 this remote settlement had its own role in the slaving triangle. There is documentation showing that the village was part of an illicit trade in slaves.

Sunderland Point is marked as the main anchorage, or outport, for Lancaster, the place where everything from sugar, ginger, cotton and tropical woods were offloaded, and also the place where the boy slave, posthumously named Sambo, was put ashore and left to die. Lancaster's celebrated Georgian architecture and the building I was researching in, all built on the back of slavery, had yet to be planned or begun.

Ulverston Canal had not yet been built, but next to the shore at Bardsea, Sea Wood was marked, once a favourite dog-walking place from my Ulverston days. I went there only a week or so after my visit to the museum. I drove over with my actor friend, Tim. He'd recently moved back to Cumbria after decades in the south, and he wanted to revisit the wood and the shore. They had lodged themselves firmly in his mind, in the way that significant places from childhood do. His mother had died recently, and Tim wanted to see the place where his family had taken him with his two brothers for picnics, and where the boys had climbed and hung from the trees like gibbons.

We walked out of the afternoon light and into the complex darkness under a tree canopy of oak, beech, hornbeam, rowan and ash. We followed the path that meandered through the wood, the bay just a shimmer of light beyond the dense branches and tree trunks. A deep limestone gully ran between our path and the shore, its steep sides preventing us from going out again into the sunlight. Eventually the slope lessened and we dropped out of the wood, emerging beside a bay that was unrecognisable from my Ulverston days.

Where once the limestone pavement had flowed down onto the sands, the changed weather and tide patterns of the past 25 or so years had reversed the process, so that now the sands flowed upwards, consuming and concealing much of the rocky clints and grikes and the pebbly shoreline. Much of it had become buried. The environment seemed to have changed so dramatically that at first Tim couldn't find the place where he and his brothers had launched themselves from tree branches onto a smooth slab of rock, and slid down the sloping surface onto the shore. We explored the edge of the wood and shore, and eventually Tim's memories began to coalesce, so that the place did form itself again, coming back from the past. We walked further and he found another location, a place where he'd made a film with his teenage pals; Tim as a gruesome clown made up with face-paint and dealing tarot cards. He's still got the film.

I was in my early thirties before I moved away from Cumbria. On a cool, late summer evening, just before I headed east, my pals and I sat with Sea Wood at our backs and we waited, talking and no doubt drinking substantial amounts whilst a foil-wrapped salmon cooked over an open fire. It was that time of the year when the season was on the cusp of change, when there are subtle shifts in the atmosphere and one begins to feel a sense of loss for the end of summer. The salmon took forever, but at least we had the bay to look out over. Since then though, forests of bulrushes have colonised the sandy shore and created a barrier between the wood and the open bay. Tim and I saw paths leading into it, places where children might have pushed through, believing it jungle. I heard the unmistakeable *tieu*, *tieu*, *tieu* and trilling of bearded tits. They were invisible in mid-plantation.

We pushed our way inside, and it did indeed become jungle, if only for a minute or two. We couldn't see the birds although they were calling tantalisingly close to us, and so we thrashed our way back out. As we did we tried to recall lines from *The African Queen* to quote to each other, but failing, resorted to 'Well quite frankly, my dear, I don't give a damn' and 'Well, I don't give a damn either!'

We walked further to a place free of bulrushes and where the view opened out again. I saw a large wooden board meshed up with weed and rope and grasses up on the high-tide line. Turning it over, there were illustrations and historical details from Cockersands Abbey and its lighthouse 15 miles across the bay, just south of the outfall of the Lune. I guessed that the sign must have become trashed by last winter's gales and been sent on a journey to the far side of the bay. There was an interpretive map too. 'You are here,' it said – and we were, and we weren't, all at the same time.

As we walked, the light changed. A new sickle moon grew out of the intense blue of the late afternoon sky and hung, limpid, above the bay. The light moved over the sands so that they turned the rough red of earthenware and the pale straw of late August grasses. A small group of people walked out from the shore. As the light flowed over the bay they were lost, found, lost and found again. At Heysham it grew dark and brooding but at Morecambe the buildings were lit by the low sun, as if they glowed from within. In the intimacy of such clear light, distances advanced and receded. We could see windows on the sea-front buildings and vehicles moving along the promenade 15 clear miles across the bay. Massive banks of clouds built white, improbable towers. Just south, amidst the darkening sky, rain fell.

* *

Michelle unfurled a paper chart dated 1890. I was beginning to feel as if I was on board ship. It showed that the channels to Grange and Ulverston were still open, still navigable. Ulverston Canal had been built a hundred years previously during the 1790s.

By now the River Lune was navigable all the way into the heart of Lancaster, and ships offloaded right outside the museum on the Georgian quayside. A lighthouse had been built at the southern end of Walney Island and another, smaller one out on Cockerham Sands. Both were intended as navigation aids for ships travelling into the River Lune and up to Glasson Dock, built a hundred years earlier.

Hoad Hill was there, along with its 40-year-old replica lighthouse, a monument erected by the townspeople in honour of Ulverston's most famous son, Sir John Barrow. Born a farm labourer's son, Barrow navigated a meteoric ascendency to the position of Second Lord of the Admiralty via a clerk's job in Liverpool, the post of teacher of mathematics in a private school, and postings to the British embassy in China and South Africa, where he was given the complex task of negotiating a sensitive truce between Boer settlers and the native black population. He was a founder of the Royal Geographical Society and held the role of Second Lord of the Admiralty for 40 years. He's remembered for sending explorers – such as John Ross, James Clark Ross and John Franklin – north to search for the notorious North-West Passage. (In one of those occasional seismic events in the world of archaeology, in September 2014 underwater archaeologists from Parks Canada and the Royal Canadian Navy's Fleet Atlantic located the wreck of *HMS Erebus*, the ship of Franklin's doomed expedition of 1845. The Inuit people had had direct encounters with members of Franklin's crew. These stories, or facts – for that is what they were – had been preserved within the Inuit oral tradition. Held in memory for 175 years, it was this testimony that proved invaluable to the locating of the ship.)

Barrow Strait in the Canadian Arctic, Point Barrow in Alaska and the city of Barrow in Alaska are all named after Sir John Barrow. As if this were not reward enough, the townspeople of Ulverston decided to build him a lighthouse; a full-scale 100-foot high replica of the Eddystone lighthouse and a landmark that can be seen from many places around the bay and beyond. It's a unique folly, a lighthouse without a light on a hilltop at the end of the shortest, deepest, straightest canal in the country.

★★

Michelle brought out the final chart, dated 1964. Just one look and it was clear that the whole bay had silted up. The individual sandbanks were melded together and the navigable channels had disappeared. Hest Bank Wharf had been built and lost again underneath the sands. In 1964, no one yet knew of its existence. Barrow was there alright, along with Vickers Shipyard and the docks where Polaris submarines were built as our first line of defence in the Cold War. My family had moved to Ulverston and my father was working at the shipyard.

At lunchtime and at four o'clock, when the yard whistle blew, a full tide of blue boiler-suited workers rode out of the shipyard gates on bicycles and surged like a wave over Michaelson Bridge and into town; you'd have to watch out if you even thought about crossing the road. The tradesmen and apprentices had a pie and a pint for lunch and then flowed back again inside the yard. Dad didn't ride a bike to work. He wore a suit and tie and travelled to Barrow by car. (And he didn't eat pies.)

No one had any idea that the Polaris missiles would stop being aimed at Russia; they would have laughed if you told them that in the shipyard. They'd've laughed too if you said that the Berlin Wall, built just three years earlier in 1961, would fall, hacked to pieces in 1989 after radical and tectonic political shifts in the Eastern Bloc. Or that pieces of the wall would be used throughout the world as monolithic, articulate evidence in art exhibitions. The word 'Taliban' had not been heard. The War on Terror might have been a story about aliens in an American comic book.

By 1964 the idea of shipping in the inner bay had become a mere historical notion. The canal, a place we were banned from ever swimming (though of course we occasionally did), and where sailing ships had waited at anchor for the fill of the tide, had been reduced to a repository for dead dogs and shopping trolleys, wrecked cars and God knows what else. The canal towpath was now just a place to walk or ride your bike down to the shore and back, passing the towers and processing plant of the drug manufacturer, Glaxo. Some days you could smell the penicillin in the air.

Given my mother's antipathy towards the bay, we didn't visit it often. Sometimes, after shopping trips to Barrow, we'd come home by the coast road, and then I saw its vast distances for myself. I could see it from my school's sixth form building too, and during art lessons I'd look up and sometimes see distant fishermen's tractors moving over the bay from one location to the next, in search of shrimps or cockles. I watched them heading back inland before the tide came in, and I remember watching its progress, altering the colour of the sands to grey as it moved along at the speed of a galloping horse.

And the hills; you can see them on the sea charts, at the northernmost part of the bay. They started at the top of our road in Ulverston. The road was brand new with plots marked out and houses half or wholly built. Me, my brother and our new friends got to work building dens on empty plots from materials left lying by the builders. From breeze blocks, bricks, wooden planking, sheets of plastic and anything else we could lay our hands on, we conjured a house of our own, using it for the whole of that summer before the builders turned their attention to the last undeveloped space at the top of the hill. In the evenings we lit it with candles that had their own place on a window shelf. If we were quiet enough, it was impossible to tell whether we were at home or not, and so for a while at least we could evade being called in for meals or bed.

We had neighbours, too, in our self-build house, a family of Californians: Captain Rust, his wife and their two boys. A handful of Americans were working in the shipyard at this time, helping the British Royal Navy to develop Polaris missiles, the raison d'être of the Polaris submarine programme.

My brother and I made friends with the Rust boys and when we called for them their mother invited us in. She was untypical of the adults we knew, and was completely unfazed by children. She asked us questions, and what's more she listened to our answers. She plied us with fresh orange juice poured from cartons that were kept in a huge refrigerator, when all we'd ever had was insipid orange squash. She cooked us waffles and drenched them in maple syrup. She frequently arrived unbidden at the den, calling out to us and handing over huge bowlfuls of home-made salted popcorn. I took to visiting her alone; she was like a magnet to me.

One day I knocked on her door as she was about to go down to the town and I asked if I could go with her, desperate for a ride in her electric blue car.

'Will your mom be okay with that?' she asked. I think she realised what my mother's response would have been as I began to slide, ever so slowly, and ever so low down in the seats as we drove past our house.

★★

My parents were both complete urbanites. Neither one of them saw the point of walking unless it involved pavements and usually shops. On trips into the countryside we were the archetypal family picnicking at the side of the road, or as near as damn it. Walks were not gone on, explorations did not take place with adults, and holidays were rare. But perhaps then, this is what makes for stronger connections, that you have to work things out for yourself. Without early signage and interpretation, we either choose to seek and see, or not. As the author Rebecca Solnit put it so well about her own parents: 'They had nothing to teach us about the countryside past the fading of the road, and perhaps the place was more vivid to me ... (a place) where no adults arrived to interpret and regulate.'[11]

So my brother and I were let loose, along with the other neighbourhood kids, to roam unregulated. We had fields and hills, trees and crags and a path that led uphill to the lighthouse without a light. Below the monument, on the face of Hoad Hill, there was a geological arrangement of stone slabs known as the Devil's Armchair. Commit a terrible deed and he'd push you over the edge with the point at the end of his tail. There was a stone gully above the largest rock face, and with our feet in this and our heads leaning over the edge we would peer down the slab where, a hundred feet below, loose rocks and the grassy slope resumed.

On summer Sundays, the folly – or Hoad Monument, to give its proper name – opened to the public and for a small amount of money you could climb the internal staircase to the top. There was just an iron handrail to hold and increasingly narrow steps to walk up, and at the top an airy platform accessed through a hatch. Only four or so people could fit at the top at once. Up there, we looked out through the open arches where, had this been a real lighthouse, the light and mirrors would have been. There was no protective glazing then, but no one ever fell to their death. With the wind whipping my hair into my eyes I looked out over the vast expanse of Morecambe Bay. I could see the length and the breadth of it; understand the scale of it. Then turn around and see the Lake District mountains ranged before me. And these landscapes, they called to me in a way that although I wasn't yet able to articulate, I felt the pull of them all.

I began to investigate maps of the Lake District, reading them as if they were books, pouring over details, exploring them in depth on wet days at home. I began to conjure stories around some of the place names that I found. The Old Man of Coniston, The Lion and the Lamb, Helm Crag, Armboth Fell, Bethecar Moor, Wetherlam, Hardknott Pass, Crinkle Crags. Going out to explore these places years later, either alone or with friends, I had to take stock, to allow them to be how they really were, along with the way I'd dreamed them up for myself; two different versions of the same reality.

★★

By now my world had expanded. I walked, frequently borrowing a neighbour's dog, a beagle that seemed impossible to tire out. Up into the hills we'd go. Passing through a field one day I found a vixen and her cubs playing in the long grass of summer. We sat quietly together, the beagle and me, his nose twitching, finding the heady scent of fox carried on the air. The cubs were just a few weeks old; they hadn't yet developed the fox's red coat and were smoky-grey in colour. Eventually the vixen led the cubs back into the den in the banking underneath a hedge. I was transfixed. That this had happened so close to where I lived.

During one of the first summers in our new home, a pair of house martins chose to make it their home too. I watched as they constructed the nest under the eaves and above our front porch. Then the chicks came and I watched again as the parent birds swooped in and out, up and down to the fields and hills that surrounded us. Returning home from any trip out, my first thought was to look up, and though I was young, maybe nine or ten, I felt that the birds nesting on our house were a small piece of magic; they'd chosen us above any of the other houses. And I knew they would return year after year.

As the summer wore on, a neat pile of droppings formed itself on the edge of the front step up to the house. It didn't ever occur to me that this was a problem, but just a fact of life. Then I came home one afternoon to find my father up a ladder, knocking the nest down with the end of a broom handle. I was speechless. His face wore an expression that told me he knew what those birds meant to me.

'Well, the chicks have gone now. They've flown the nest. Your mother doesn't want them coming back again – all this mess to clear up.' He indicated the pile of droppings. A chasm had opened up between us. I'm not sure that it ever fully closed again.

* *

Later, in the 1970s, the shipyard had a new wave of foreign partners. Warships were being fitted out for the Argentinian navy. My mum and dad became friends with a group of the Argentinians and their wives. They came to our house for supper and my parents went to theirs and together they went to pubs to listen to Trad Jazz, or as me and my musician pals later called it, 'shipwreck music' (every man for himself). In just another few years, Argentinian ships and ours were blowing each other sky high out of the South Atlantic Ocean in the Falklands War.

Seven

Silverdale

GARLIC TANG ON THE BACK OF YOUR NOSE. The wide floor of the wood was greening up, intense viridian. Leaves pushing upwards, migrating toward the light, each single one a scimitar arc, the tip nodding back down towards the earth. In a short number of days the woods would be full of the white globe stars of ramsons (wild garlic).

The sign, leaning slightly, pointed us through the woods and towards the shore. The footpath began as a lane, bordered by mossy drystone walls. The Silverdale area is riddled with paths that wind through its ancient woodlands of hazel, sycamore, beech and sweet chestnut, ancient ways that forged crucial links between one small community and another. Silverdale, from the Norse 'Silure's Dale', is a place of clearings and woodland, enclosures and wild places side by side. A random patchwork of fields cleared where the rocky outcrops allowed, woodland for field borders and margins. Tree roots plunder the clints and grikes of great bluffs of limestone. Pathways forge themselves between and around the silvered rocks that give these woods their character.

A woodpecker drilled pneumatic, unseen. Diffuse sunlight measured the spaces, bands and beams of light came in, trees just beginning to show their colours again on the back of our longest winter for years. The pungent scent of wild garlic filled the air and filled us up; it was impossible to breathe and not take it in. It bordered the wall bottoms and spread in drifts, claiming the whole woodland. Just a couple of weeks ago the woods were in thrall to bluebells, casting their own particular spell. In Beckmickle Ing, near our home bordering the River Kent, and in Dorothy Farrer's and Mike's Wood too, we'd go frequently to take in the evening magic of them, the low sunlight slanting into the woods and the cobalt distances receding towards the woodland horizon.

We went to Silverdale *en famille* – my husband Steve, son Fergus and Milly the dog. She was alive to the scents of the wood, following and chasing lines of communication left long ago by foxes and rabbits and voles. Other dogs too. But there was something else also, and Milly knew it. I could tell by the set of her eyes and the way she charged about, head extended. Then there it was, bounding and looping across the ground and up into a tall beech. A grey squirrel that gained the upper storey in seconds and ran along a branch, then off into another tree and away into the intertwining canopy. All Milly could do was to stand with her paws as high up the trunk as she could manage, trying to work out, as she always does, exactly where the squirrel has gone.

At the bottom of our lane in Kendal there's an old yew tree extending a gabled branch out over the tarmac. In snow the lane is impassable by all but sledges and boots. One New Year we'd had a fall of the most crystalline snow I've encountered. The flakes fell and froze as the temperature dropped and stayed down for days. I'd walked down the lane with Fergus and the dog on our way to the Green, Fergus fiddling about with something, I can't remember what, but he was behind me, travelling in my snowy wake. Milly got the scent of a squirrel. It crossed the lane in a grey flash, then up onto the trunk of the yew. From my position further down the lane, I turned to watch and laugh at the dog adopting the squirrel-hunting position, paws up on the bark.

The sun had just risen over the fell back up beyond the house and was still low to the sky. As the squirrel ran out along the broad branch over the lane, it dislodged a shower of frozen crystals. I watched the sun catching each super-lit particle as it fell, turning them celestial, and into it all, unaware of the effect, Fergus walked, so that he became lit with the sparkle and shimmer of it all. It was an extraordinary moment, this falling of fire-lit snow and my son caught in its magic. The dog oblivious, still nosing skywards, bemused.

* *

My husband Steve lived in Silverdale when we met, and a pattern of weekend visits to my home in the North-East or to his in Silverdale evolved; best of both worlds. It wasn't an area I knew very well, and it was good to walk the mazy network of paths with an expert, setting out from the front door with two dogs, though frequently returning home again with only one. The area is home to every deer species in the UK, unsurprisingly, as the woven landscape gives them places to feed and leap back into the wood in seconds. Steve's dog, Kim, a golden retriever-Labrador, was incapable of resisting the scent of deer. We'd go walking and then the dog would shoot off, deaf to our calls to come back. She ran and ran until exhaustion eventually overtook her, then she'd come back to the road and always managed to find some hapless dog-lover. We'd reach home and later, there'd be a phone call.

'I've just found your dog.'

And Steve would reply, 'OK, that's great, thank you. I'll come and get her. Where is she this time?'

The path gently descended towards the saltmarsh at the edge of the bay. A series of creeks act as drainage channels wicking excess water away from the pools of Leighton Moss RSPB reserve, and a signpost points to Quaker Pool, the footpath running along the top of flood defence banking. Beyond it are fields and further still the ochre swathes of bulrushes at the reserve. Further away towards the east the sloping and folded strata of Farleton Fell. The Furness railway line passes close and, as if spirited out of nowhere, a train came, horn blaring a warning for the level crossing ahead.

We separated. I headed off to the newer hides of the Allen Pools and Steve and Fergus set off to walk the dog around the coast towards Jack Scout. They had to round the small, low headland at Jenny Brown's Point, where there's a strange intrusion in the landscape. It's a lime kiln chimney, all that's left over from another era and another view of how the landscape was used. In the 1790s a copper-smelting works was built there, laboured by migrant Welshmen at a time when copper was in great demand for the manufacture of cannons to feed the Napoleonic wars. The machinery and boiler for the works were brought in by boat to a purpose-built quayside, and then removed again by boat when the industry declined. All that remains of the quay is a toppled pile of limestone blocks. The chimney itself is an anachronism, but I think it fits; it seems to belong. It puts me in mind of the Mannerist landscapes that followed the High Renaissance. How were artists able to move on from the perceived idea of perfection? They began to exaggerate, to create visual puzzles and odd perspectives. Here, the visual puzzle of the lime kiln chimney that speaks of another era.

I turned to look back at the woods we'd walked through. They were still so brown, as if the life has been singed out of them by this ridiculous measure of cold weather. Looking harder though, and I thought I saw just a rebellious tinge of pale green; spring moving in at long last.

From the top of the bank and just above sea level, the bay itself was a thin, silver band pegged into place by the Heysham Power stations and the hazy line of Morecambe's sea-front buildings.

On the journey around the liminal spaces of the bay, I learned about orientation, about how the bay comes at you in unexpected angles and how you see places from a new perspective. A swathe of saltmarsh swelled out towards the tidal zone and sheep grazed the surface. Pairs of shelduck rested, occasionally ruffling feathers and settling again.

A single swallow skimmed the air – my first of the year. I walked underneath a hazel tree, its emergent leaves not yet fully unwound and hanging in pairs, like insect wings held out to dry. Coming closer to the Allen Pools the clamour of black-headed gulls was unmistakeable, raucous, the sound travelling over the low-growing willow and tall grasses that gave shelter to the pools.

I had heard that avocets had come in to Leighton Moss and that their numbers had been increasing, but going into the first hide, I was completely unprepared for the sight from the windows. Wading, probing, investigating the underworld of water, resting on innumerate small islands or checking out the airwaves were some 50 or 60 avocets, stilt-legged and distinctive; a black and white spectacle of birds. Some of them were so near to the hide and the open window that I could hear the quiet plash of bird-foot rising out of and entering back into the water; in and out, up and down. Taking the time to tune in, and tuning out the squawks and squabbles of the gulls, there was the sweeter, more simple and quieter piping of the avocets.

Avocets were once relatively common in many areas of the UK, but with significant losses in their wetland habitats, by 1840 they had died out completely. Their rebirth as a breeding species was down to the flooding of the coastal margins of East Anglia during the Second World War. Gradually, a very few birds returned to British soil. In May 1997, a single avocet was seen on the saltmarsh close to Leighton Moss, a single bird that must have had the twitchers jumping. Five years later a pair bred, an even more remarkable event given that Morecambe Bay was considered far north of their typical breeding grounds. The first nest failed, but the birds persevered and two chicks hatched that season, one fledging.

Slowly, more birds began to arrive, though it took years for their numbers to grow. In 2011, nine young fledged, but in 2012, some 48 young fledged. Well over a hundred avocets have now been added to the UK population through Leighton Moss.

Previously I'd seen avocets only once before. Arriving at Minsmere in Suffolk whilst on holiday one August, the warden's answer about the best place to see them was not what I'd anticipated.

'I think you're a bit too late. They've already gone.'

Walking the circumference of the reserve, each hide drew a blank. Then at the final hide a man looked up from his telescope, and smiling he said, 'You want to have a look?' and there in the sights, avocets. It was brilliant seeing them for the first time, and sharing the man's similar delight in these semi-exotic creatures. But to see *all* these avocets here by Morecambe Bay, a world of them, was just great.

Eventually I tore myself away. Leaving the hide I sensed a movement overhead, and looking upwards a marsh harrier was passing over the pools, mobbed by a single oystercatcher. They flew off together. The harrier, a male, its undercarriage hazelnut brown and with paler wings, banked and dropped over the density of the reed beds, tumbling as if weightless in the air and giving an occasional one-note, cat-like call. The harriers are mostly migrant birds that return from wintering in Africa during the spring. Occasionally single birds have been spotted during the winter months, and it's thought that some individuals stay put because of the milder winter climates of more recent years. Even though that particular spring had decided to be later than ever, the birds had returned to Leighton regardless. They were already on the look-out, and being looked-out for.

There must be something in the air at Leighton Moss; the much rarer raptor, the hen harrier, is seen occasionally flying over the pools and reed beds. These birds have been subject to persecution on an unprecedented scale. Perhaps as few as three pairs remain in England, though they fare much better in Scotland. In England hen harriers breed on areas of open moorland that are used for shooting grouse, and mysteriously, whenever a brood hatches, the young don't seem to survive. Sometimes the adult birds are shot dead. But there's a tide beginning to turn, and perhaps the raising of the persecution of hen harriers in a wave of press articles, together with peaceful demonstrations close to the birds' habitat, may eventually bring about change.

★★

A lone lapwing wavered across the sky and out towards the bay, joined by another, and both disappeared over the reed beds. The hides were quiet that day. In the second, just one man sat looking out, his telescope trained on the dark forms of hundreds of birds on the lagoon. I'm admittedly rubbish at identifying waders and I asked him what species they were.

'They're black-tailed godwits. They should really be gone by now.'

'Where to?' I asked.

'These birds from Morecambe Bay head up to Iceland for the summer, but I've never known them stay this long. It's this winter, it's just going on and on.'

I told him I'd seen my first swallow, and that it must be spring soon, surely?

Coming in through the open windows of the hide, the call of the godwits had the distinctive sound of the marsh about them, of liminal places, a lilting, wistful repetitive tune of three short notes. More lapwings wafted above the pool and out towards the saltmarsh, *pee-witt*ing as they travelled. The godwits had settled in groups, mostly sitting tight, either in the water or on the small islands. Most had their heads turned and tucked into their back feathers, and those that hadn't were aligned with their heads to the north-west. With the weather apparently settling down, I wondered how long it would be before they left, and I looked up to a sky that was a pale, washed blue layered with milky bands of high cloud. Occasionally a bird lifted from the water, skimming the surface like a moth, white wing-bars flashing before settling again. The birdwatcher packed up his kit and moved on. A moment later I followed and as I walked down the wooden ramp a bearded tit landed on a fencepost a few feet ahead of me, its feet planted wide. I stopped stock-still to take in the orange plumage, the head the same colour as the sky, the white chin and long tail feathers. Two dark curving patches below the eyes, like heavy mascara that has bled in the rain, or a heavily made-up Japanese kabuki actor. One of those eyes turned towards me, head to one side, then he flashed away into the high hedgerow, and though I could no longer see him, he poured forth his song: *sweep churrrrr, sweep sweep churrrr, sweep churrrr.*

★★

Out on the thin sliver of the bay a distant heat haze had risen and, seen through it, the hills of Furness broke up and were pixellated into indeterminate shapes. Above, the inevitable vapour trails melted into nothing. Clouds were beginning to break up into small shapes like cats' paws on the surface of a lake. Buds were forming on willow and, seen from a short distance, were like lime mist. Another songbird, a carmine-chested bullfinch, chirped and whistled as I passed below. The first insect hum of spring evolved from underneath the tree canopy. Swallows came, gliding and hunting for airborne morsels. One swept low and passed in front of my face. It almost touched the end of my nose.

Further along the path there's a barn and a hole in a gable wall where swallows dived in and out, chirping. I closed my eyes and as the warmth of the day pulsed and grew I sensed that spring was here, and that the trees were pushing out leaves responding to the shift of season – the 'green fuse' of Dylan Thomas.[12]

I walked back along the bank and saw Steve and Fergus walking towards me, rounding the point by the chimney, the dog bounding on ahead. I called to her and she came running flat out along the path, skittering to a halt and then setting off back towards the others. It's the collie in her; she likes to keep us rounded up.

I looked out at the edge of the bay and there, crossing the saltmarsh, heading away from the pools and then following the line of the land-edge, a large flock of waders were winging towards the north. Too far away to identify clearly, but I fancied they were the black-tailed godwits, released by warmth; as if they too felt the green fuse and knew the time was right.

* *

When our firstborn son, Callum, was two weeks old, we left my house in the North-East and came to Silverdale so that the new arrival could be duly shown off to family and friends. He was born at the end of July and we were blessed with good weather. As I slept on in the mornings Steve wrapped Callum into the sling and took him with the dogs for a walk and to fetch the paper from the shop. We walked every day, introducing the boy to the quiet beauty of the Silverdale woods and shore.

We'd walk along the road, passing the Lindeth Tower house in the grounds of a high-walled garden. The writer Elizabeth Gaskell used the three-storey tower in the 1840s and '50s when she escaped the Manchester smog for recuperative holidays. She wrote her novel *Ruth* in the newly built tower, and her short story *The Sexton's Hero* is set on the bay (see Chapter 5, 'David Cox: The Road Across the Sands'). In 1858 she wrote to a friend that she was due to visit Silverdale for a six-week trip. 'The house,' she said, 'is covered with roses, and great white Virgin-sceptred lilies and sweetbriar bushes grow in the small flagged square court. At the end of the garden is a high terrace at the top of the broad stone wall, looking down on the bay...'[13]

And Silverdale is where Charlotte and Emily Brontë spent just one night, staying in The Cove, a house owned by the Reverend Carus Wilson, headmaster of the girl's school at Cowan Bridge, over the border in a north-eastern corner of Lancashire. This school with its diabolical regime, and where the two older Brontë girls contracted and later died from tuberculosis, was reconstructed as the unforgettable Lowood School in Charlotte's *Jane Eyre*. It was mindboggling to read that the girl's father continued to send them to the same school after the death of their sisters.

With our new arrival we'd walk on towards the limestone headland of Jack Scout and stay a while on the Giant's Seat to look out over the wide bay. We'd take in the view, the tide in or out or travelling, the headland further north with trees sculpted by the wind, the branches of each one rolling over and intertwined within the branches of the next. Oystercatchers and curlew gathered on the sands below, or called as they passed, skirting the cliffs below us on the wing. In the scrubby vegetation of the headland, blackbirds surfed the air from one hawthorn to the next. I have photographs of us walking home in the sunset, each of us smiling – always – the baby carrier strapped to our front and the evening sun turning the silver-grey limestone cliffs rosy.

The holiday became a procession of visitors. Steve's parents came, the wonderful John and Peggy, finding themselves doting grandparents for the first time at a fairly advanced age. Our friends from Ulverston came as a gang, kids and parents bearing gifts for the baby and alcohol to wet the baby's head. We flew kites on the shore with the children and talked and laughed. Steve's stepdaughters from his first marriage came, Anna just returned from New Zealand, and Eleanor living down the road in Lancaster. All this evident pleasure for us and the new arrival was truly moving.

I am writing of a period when time seemed suspended. Not merely through the broken nights of a baby's demands, but because my memory tells me that we lived by the light. It called us out in the day and evenings and we went to bed with the late sinking of the sun. I learned the network of lanes and trods and footpaths and discovered the idiosyncratic wells of Silverdale – Woodwell, Elmslack, Burton, Bank and The Row. Walking the footpaths then, and looking down at our new child with the sunlight and shadows from the woods passing over his downy head, and thinking now of those days, what blessed times.

In a previous year, on one of my first visits to Steve's house on Lindeth Road, I went walking one morning with Kath, the friend who'd introduced us. Along one of the village's hidden pedestrian arteries was a cottage set in a walled orchard with a hand-painted wooden sign that we'd seen the day before: 'Plums for Sale'. We walked towards the front door, which, like all the windows, stood open, curtains from another era blowing in the breeze. The walls of the building were faded and yellow like an ancient Normandy farmhouse. Loud, transistorised classical music poured out from the open ground floor windows and door. We knocked, but there was no reply. After shouting hello and knocking again, we peered inside and saw, there in the kitchen, his back towards us, the old man whose plum trees groaned in the orchard under the season's weight of red and yellow Victoria's. We watched the man, his arms moving through the warm air as summer flies droned in bozzy patterns above his head. He was conducting music coming from a small radio on the kitchen windowsill.

We knocked again and tried shouting louder, but eventually Kath walked closer and called again during a lull in the music. The man was, unsurprisingly, a little taken aback, but then seemed pleased to see us.

'I rarely hear people knock first time,' he said. 'I'm very deaf.'

Once we'd made our request and he'd understood, he took folded brown paper bags from a drawer and filled them brimful with plums, taking hardly anything in exchange. We talked for a while, us two speaking clearly, and then took our leave. Walking back to Steve's, I thought of the man, struggling to hear much at all but still finding an emotional connection with the music, or even with the memory of the music. I thought too of his pared-down life, of an interior world that belonged to another time.

Years later Steve and I walked that way again. The cottage and orchard had been bulldozed, replaced by a much larger house and a garage the size of the original building, four-wheel drives outside on the gravel. The garden was all raised beds and trellises and archways in new timber. I wondered then if the memory was real, or if the man had been woven from magic and helped me to fall in love, not just with Steve, but with Silverdale too.

Eight

Foulshaw – A Lost History of Peat

I WENT LOOKING for the place where the sea comes closest to my home. I wanted to witness the outfall of the River Kent into the bay. The river makes its final meanderings close to the magical topiary gardens of Levens Hall and on between flood embankments at Sampool. Here the Kent makes a final series of turns that cut the land into jigsaw pieces before it begins to widen only a wing beat away from the sands and finally pours itself out to the wider world.

From these utterly flat lands the river's birthplace, the mountains of Kentmere, form a distant backdrop. On that March morning the tops were smoothed over, white with even more fresh snow. The wind was a blast from the north and icy cold, bringing tears to the eyes as I walked towards the bay along the Foulshaw road.

A small group of swans were feeding in a field and a pair of crows took to the air, moving down the lane ahead of me like gothic tour guides. They flew lazily, keeping within the confines of the black hawthorn hedgerows, the light silvering their backs with an oily gleam. One after the other they peeled away, leaving me a solitary figure walking down the road.

I followed the direction of a footpath sign for the Cumbria Way across fields and at last the path led under the protection of the embankment. Thank God – that wind was the fiercest and coldest I'd felt for years. There was a feeling of walking in a sunken lane, the bank is the width of seven or eight men and was just high enough to stop me from seeing anything beyond it.

Too soon the bank peeled away, swinging across the field edge. I knew that on its far side the River Kent entered the bay, but there was no alternative but to keep to the path alongside substantial fencing. Large pools of standing water in field depressions were coloured silver and splashed with wintry blue. A scattering of sheep cropped at the grass oblivious to the cold. I walked past a line of ancient hawthorns, maybe a quarter of a mile in length. Their erratic growth patterns must mirror how the wind comes from all directions here. Strings of sheep's wool garlanded their lower branches.

The bank swung back again across the field to rejoin the path. I passed through a small gate and at last I was able to climb onto the top of the bank to see the meeting of river and bay. The extreme blast of freezing wind that greeted me, and that battered me around the head and face, brought more blurry tears.

But there was the marsh, dotted with skeletal winter-black gorse bushes, and with the tide far out, the channel and path of the Kent was clear, moving towards Sandside and Arnside and the south, and surrounded by the ochre distances of the sands. The sun moved behind bands of high cloud and the light became suddenly monochrome, making the farmland and marsh almost devoid of colour. I saw a red post van travelling up the lane from Low Foulshaw, the only coloured object in the whole landscape. There were sheep out on the marsh; I envied them their apparent obliviousness to the cold. On the edge of the channel a pair of mallard was silhouetted. Two more took off from a field and flew towards the bay with steady wing-beats. They passed overhead and I could see the way their bodies articulated, hinging slightly up and down as they moved through the freezing air.

The Kent flowed on towards Arnside and the railway viaduct, its piers and gaps in turns blocking and revealing the steel-coloured path of the river beyond. Two walkers, black against the light, appeared in the distance walking along the top of the bank; at least they had the wind behind them. They were walking away from the next tiny coastal settlement of Ulpha, named from the Old Norse for wolf. I could imagine this landscape being roamed by wolves now – good pickings with the sheep on the marsh. At Ulpha there's just one farm backed by a hilly, wooded outcrop and closer to the coast a single cottage.

I walked further, but eventually, defeated by the cold, I moved down into the shelter of the bank again. A train blew a warning *haw-hee* as it approached the viaduct. A drainage channel ran in confluence with the bank, the water in it like treacle, and hawthorn branches hung over the path where sheltered sheep haunts were revealed by tatters of mossy wool. I found one white shoe that had become subsumed by the grass; it grew inside and around it. I wondered about its provenance: why one shoe? And why here, on the sheltered side of the bank? It looked far too new, too clean to have been washed up by the tide.

From the far side of the bank I heard the distinctive call of geese in flight. They appeared, winging low; pink-foots, whooshing right overhead. I counted 22. The sound of geese, *and* the sight of them, has always been a complete symbol of the wild for me. In Ulverston, I'd heard them passing high overheard in skeins. Sometimes I'd be in bed at night and the sound came travelling in through the open window through the pitch dark. When that call comes in over the airwaves, I have to stop whatever I'm doing and, with a slight pitch in the heart, I'll race to the window and look upwards; the small miracle of geese.

* *

As I walked back towards the fields, there came from the far side of the bank the unmistakeable sound of curlew, but it was magnified many times and intensified rapidly. A second later a shape-shifting flock of 50 or even 60 curlew materialised above the embankment, rising skyward with a shimmer of movement. Their marled colouring feathered the sky. Mirage-like they pulsed against grey clouds. I'd never seen so many in one place before; it was an extraordinary sight. I watched them, revelling in them. Moments later a small flock of starlings flowed into view behind the curlew. The movement amongst the curlew was one of agitation; they rose and fell in wave form.

Out of nowhere a peregrine dropped out of the sky like a dark stone. It pierced the starling flock, and with the acuity of a marksman it picked out a bird. I saw the starling struggling momentarily but within seconds any movement had stopped.

The panicked starling flock dropped as one towards the ground then streamed low, skimming the fields towards a sheltering line of Scots pines, and the curlew retreated again behind the bank. The peregrine seemed to be making heavy weather of its kill, struggling to keep height and direction of travel. Then two tiny birds, no bigger than goldfinches, flew after the hawk and began to mob and harass it. The bigger bird struggled. It strove to move away, eventually coming to land in a bank of hawthorn. My freezing morning had struck gold.

As I walked back up the road, a flock of chaffinches took flight from the hedgerow like a handful of pink and grey leaves snatched away by the wind.

★★

I came to Foulshaw again a week or so later, but this time to Foulshaw Moss, a vast area that was in the process of being restored back to its original manifestation as blanket bog.

You can see the moss from miles away. Even from the distant vantage point of Scout Scar near Kendal, Foulshaw Moss stands out, appearing like a vast pale oasis amidst all the green of fields and woodland at the edge of the bay.

I'd gone with my friend Brian Fereday, whose family was one of the last to carry on cutting peat on the moss until recent years. It was as much a part of their lives as the growing of vegetables. As we walked, he began to orientate himself, remembering the area of ground they'd last worked.

We walked onto the moss through a gap in a line of scrubby birch trees that bordered the lane, then out onto a semi-wetland, last year's grass as bleached by the weather as old straw. The ground was hummocky and rough, with pools of standing water.

'I'm pretty sure it was here. Yes, it was. They're letting it flood again now, no drainage happening any more. You see the line of that bank? That was the working edge, the face of the peat, and the spreading ground behind it, where you left the cut peats to dry out.' I saw features beginning to emerge.

'Traditionally the men cut the peat and the women turned it; you had to keep turning it until a skin formed.

'In the past the big landowners had control of the cuttings, going back centuries. Families paid annual rent for a "moss room" – great name, eh? It's all documented, and each peat cutter had to keep their part of the moss drained to keep it dry. There was a main drain down the centre and side drains – all run by gravity and sluice gates. There's one that still works now, just over the far end of the moss, at Ulpha.' And Brian pointed out towards the bay.

I looked out across the huge space backed by a broken line of shelter-belt pines and here and there single trees that had been cut off, I guessed, at three-quarters of their original height. It seemed more like the aftermath of a war zone than a nature reserve, but still I half expected a marsh harrier, or some other raptor, to hove into sight – the broken trees seemed ideal perches for hunting.

'To come here to cut peat, you had to make an effort. That's a good principle,' Brian said. 'You'd take out what you needed. No more.

'It used to be a big place for gulls to nest, lesser black-backed mainly, and people came gull-egging. I never fancied eating them myself though. There's curlew too; in late spring dozens of them nest out on the drier ground. In the past the landowners used it for shooting, wildfowling. If you had a dog and let it run loose and disturb the birds, the dog could be destroyed – if they found out. We're talking 1700s here. But in our time, in the 1960s, the moss was planted up by the Forestry Commission. They filled the place with pines and spruce but they never came to much. This ground's too wet for trees so the timber was poor – you'd think they'd have known that.'

Many of the shelter-belt trees around the edges of the moss appeared drunken, tilting at angles, leaning into one another as if for support.

'They don't look too happy now,' I said, indicating towards them.

'Aye – what's left will fall eventually, but that'll set up a new environment of decaying timber. What goes around, eh?'

Under foot the moss banks were bouncy, springy to walk on.

'We'll go and have a look at the boardwalk.'

We stepped up onto the higher ground of the bank and close by a snipe rose out of a drainage channel, jigging away, following the line of trees by the path. My feet squelched through the bog and a tideline formed on my boots. I could see the line of the bank following the birches back up towards the main road.

'The best peat came from further down in the ground, where it'd been under greater pressure; on its way to being coal. If you dropped it on hard ground it would break apart like china. That was the best.'

As we walked, my eyes continued to adjust to the landscape as it had been, making out features: drainage channels, pools and peat edges, now a maze of banks and water.

'To my untrained eyes it all looks primeval, as if it had never been managed… well, apart from the broken trees,' I said.

'But what man makes, eventually will be unmade. This new regime, they call it "re-wilding" nowadays, don't they? They'll restore it back to the original landscape, though there are plenty of locals who don't see why you can't have both.'

A pair of ravens passed close, lazy wing-beats carrying them off towards the bay. We stopped to look down at a pool full of frogspawn, frozen, dead to the world.

'This winter; it's never-ending.'

We came to an old section of boardwalk, the first to be built when the site was taken on by the Cumbria Wildlife Trust. But as the level of the water in the moss increases with the blocking of drains, so the boards, like the ground itself, rose and sank and in places the cambered edges were lapped at by black, peaty bog water. This part of the moss was a different ecology: stands of young birches were being established (get the birches in first and the rest will follow). Scot's pines a way off, scrub willow, bog myrtle and dark purple heathers. We heard the high-pitched *tsee, tsee, tsee* of long-tailed tits and eventually they broke cover, rising into the air and filtering into the top of a birch.

Near the end of the walkway we came to a simple square-built shed. Brian told me that each family had had a hut like this, a place to keep a peat barrow and spades, and to store peats for a year or so to carry on drying out.

We peered inside and there, abandoned, like an object in a museum, a peat barrow. It was a primitive hand-built affair with wooden handles, a metal frame and a large-diameter wheel sinking into the compacted floor of the shed. A wooden board was fixed to the underneath of the supports, and in the barrow itself, a few pieces of peat, broken, exposing their crumbly, dark interior.

'A good place for insects to hole up,' Brian said, and he picked up a piece of the peat and broke it into two halves, a dusty residue minutely clouding the air around it. More pieces littered the earth floor.

'See the wooden board? That was to stop the barrow sinking into the ground when it was fully loaded.' Then he seemed pensive, and said, 'I didn't know this was still here. Thought it'd gone with all the others. In two or three years the whole place has been cleared of all this. It was a way of life, gone forever now though.'

And I thought that if it hadn't been for this last piece of evidence, I'd not have fully understood. The ground of the old 'moss room' was in transition; it hadn't fully become its new re-naturalised self and just enough of the old remained to see it as part of a rich historic seam. The evidence was there on the ground – once you'd got your eye in.

* *

Skip forward to the back end of summer a year later. At Foulshaw the collapsing boardwalk had gone, the peat cutters' hut too. There was a swanky new boardwalk out into the moss, and a viewing place where you might have the luck to spot an osprey. This season three chicks had fledged from a nest built in the top of a Scot's pine. On the day I went to meet the reserve warden, a single chick remained in the nest. The parent birds had both already set off on the migration back to Senegal, the female leaving shortly after the chicks had fledged, and the male some time later. He has the job of staying to look after the kids, feeding them and teaching them until it's time for him to go too. Then finally the chicks set off, each one travelling separately. There's a yellow sign at the roadside now, 'Foulshaw Ospreys', and a new, clean interpretation shed, new gravel for car tyres to crunch over.

'Ospreys have been coming here since 2008,' the reserve warden, John Dunbavin, told me. 'But this is the first time they've bred successfully. When they first came in, they were immature birds, offspring from other breeding sites in the area. The whole thing's documented on our nest camera. It's on the website too with a blog people can follow.

'They're faithful to their nest site, so if we can attract them in using ready-built nest platforms, and if they breed successfully, the chances are they'll return year on year.'

The Cumbrian osprey story started further north in Bassenwthwaite with the first birds arriving in 2001 after an absence of 150 years. They're well established there now, and the young birds are spreading their wings, returning to England from migration to find new territory in Kielder Water in Northumberland, in Esthwaite and now here beside the bay. I'd heard talk last summer of ospreys seen fishing over the bay, and this place, well, the larder's a mere wing beat or two away.

'They're amazing birds,' John said. 'Well, they *all* are; the whole business of migration. They have a genetic magnet inside their heads.' I liked the image.

'We've had some fantastic sightings of them, flying overhead with fish in their talons. They really are the icing on the cake for us.' As John talked, I peered through his telescope at the last chick, though it seemed fully grown. Immobile, dare I say statuesque, with its dark, hooked bill, yellow eye and white chest colourings.

'It'll be off any time now, any day. Then its fingers crossed for next year.'

We turned our focus to the ground itself. I told John of being here with Brian, and about the changing order of things.

'It's impossible to keep everyone happy. Whatever you do, someone's not going to like it. Some of the locals have complained, saying that blocking off the drains is causing flooding further up the valley, but it's just not possible. You have to have an eye on the future. A big peat bog like this is a carbon sink. Peat bogs trap carbon, which is a necessary thing these days. It works globally, not just locally.

'Before we started work on the site, the peat bog was drying out, and if it carried on drying out it wouldn't be able to do its job of storing carbon. Now it's a Site of Special Scientific Interest. It's a Natura 2000 site too, designated as important for habitats, important nationally and internationally.

'When I first started here, you could walk across the whole moss in trainers. You couldn't attempt that now. In terms of wildlife we've got dragonflies, frogs, birds. We've had a hobby hunting down here in the evenings, feeding on swallows and house martins. And it's a supermarket for bats down here.'

I made a mental note to come back much further into the spring and on a warm evening.

'Burning peat though, it's just not sustainable; digging it up releases carbon, and if you burn it, it releases more. People are more aware these days, certainly about peat from England and Ireland, but the problem just moves further out. It's shifted to Eastern Europe now, where it's being dug out in vast quantities.'

Although it was just over a year since I'd last been to Foulshaw, there was a palpable change in the landscape. The water levels had risen; it would be impossible to set off to walk anywhere now, even across the old peat cuttings that Brian had brought to life again.

'We've been draining wetlands for agriculture since Roman times,' John said. 'In this country alone we've lost 96% of raised lowland bogs, so it's really important to restore it. At one time this whole area at the northern end of the bay was one continuous peat bog: Foulshaw Moss, Meathop Moss, Nichols Moss, Stakes Moss and the Winster Valley.'

I knew that this was why the cross-sands route was used until the railway came; the only other route involved a tortuous and lengthy detour over high roads and steep inclines over the fells to the north of the moss.

'Over the centuries the mosses were gradually drained. Even though it seems large, and it is, Foulshaw is really an isolated fragment. It's all that's left.'

Like a relic from the past that's been buried, smothered by vegetation and released again into the world. I left John with a sense of the imperative of not only managing this site, but of the importance – no, crucial need – for active conservation. Just after my visit I'd heard a scientist on the radio talking about his lifelong study of bees: 'I don't get what's not to like about conservation. What's the matter with politicians? Don't they get it? Conservation *is* the future.'

I took a final look back at the nest. The osprey was standing up, beating its wings and stretching them out. I felt, if I stayed on, I might just see it leave for Africa.

★★

There's another story about being buried and revealed again, something that would've remained undiscovered had peat cutting not been such an integral part of life here. Reading the literature on the archaeology of Morecambe Bay, it's thought that early man's occupation of the area was patchy at best. Of course there are finds and evidence, but this story is about something wrought out of collective and purposeful action by a sizeable community.

In 1897 peat cutters began work on a previously long undisturbed and isolated area of Stakes Moss, lying to the north of Foulshaw. They discovered a 'trackway' and the find has been described as 'the most impressive archaeological structure recorded from the South Cumbrian wetlands'.[14]

An eyewitness, J.A. Barnes, described what the peat cutters found:

> *Cross timbers laid side by side on three lines of supporting logs parallel to the direction of the road. The larger timbers, some of them 2 feet 6 inches thick, had been split and laid face downwards; the smaller ones were left entire. At short intervals along each side of the road, pointed stakes (perhaps the origin of the name Stakes Moss), had been driven deep into the peat to keep the supports from slipping outwards. We dug out one intact, and the point was 3 feet below the level of the road. No nails whether of iron or wood, were observed. The material is mostly birch… When first got out it has almost the consistency and appearance of gingerbread, and may be cut easily with a spade, but it shrinks and hardens when exposed to the air. It is not rotten, but, as it were, pickled in the juices of the peat.*[15]

I love this description; I'm almost there with them, smelling the acidy earth and seeing the centuries-old timbers revealed, 'pickled in the juices of the peat'. But as with the raising of the Marie Rose, the sudden change of environment and atmosphere induced chemical changes immediately upon contact with air.

The diggers saw distinct axe-markings on many of the logs, and from the marks they reasoned the cutting edge of the axe to be one and a half inches across, perhaps a 'finishing' axe. No doubt the axes used to fell the timber to make the trackway were of a larger configuration. In total, the section revealed was 15 to 16 feet wide and 50 yards long.

Half a mile south, another section was discovered, but here the trackway ran continuously for 180 yards. Barnes wrote:

> *I remember when quite a child hearing mention made of a wooden road buried in the peat moss, and when the piece before us was discovered I imagined it was the one I had heard of before.*
>
> *But on making further enquiries I found they were wide apart, and after considerable grovelling in bramble-grown ditches I discovered traces of the other piece on both edges of the moss... [If the two sections intersected] it would be at an angle of perhaps 60° near the River Gilpin.*
>
> *They were probably branches of a single road from the east which forked after crossing the [River] Gilpin, one branch going north towards Lyth and Bowness, the other towards Witherslack and Furness.*[16]

So here was evidence of early man planning and building a trackway to help them traverse the moss and take them to the river. Their settlements were likely to have been higher up on the flanks of Whitbarrow and Helsington, but here, at the intersection of their road, was the river – an obvious source of food. The trackway, and the people who made it, have been dated to mid-Bronze Age, 1550 to 1250 BC.

I like too that idea of Barnes', that there were rumours, other sightings if you like, from the past. The idea that trackways had been glimpsed before and recorded by nothing more sophisticated than the telling of stories. And all those layers of time; the laying down and combining of time and moss and water and the formation of peat over centuries. All that continuous, unobserved chemistry.

Back into the 21st century, if the mosses are to remain protected there's little chance of further archaeological surveying. Until the arrival of sophisticated new detection equipment, other evidence will stay hidden. And as time moves further into the future, any talk of roadways underneath the peat will become just another local rumour, a story to tell in the pub or to the grandchildren, something passed down the generations.

Nine

The Drowned Wood and the Boat with a Wood in its Heart

I'M WALKING THROUGH A WOOD and it's slowly filling up with water. It must be spring or autumn as light falls through the tree canopy. I'm paddling through clouds and at the same time I can see stones and gravel, plant life, tree roots. I can't tell from which direction the water is coming. And is it the sea or the river? It's a dream sequence, and it isn't.

* *

I was walking with a friend and our dogs. We'd heard good things about Roudsea Woods, but we hadn't known this. We parked by the Haverthwaite Road and walked in with the River Leven at our side. It flowed deep and dark and fast that day. I remember thinking that the dogs wouldn't stand a chance if we threw sticks in for them to fetch, and I was extra wary as this had happened to my dog Molly, and in a small field stream at that. I'd plunged in after her. I was alone, and feelings of surprise and shock and stupidity followed as soon as I'd waded in and felt the strength of the flow against my legs, nearly knocking me off my feet. By the time I'd sorted myself out she was free, having been swept across to the far bank where she clambered out and shook herself dry. Then I'd had to squelch home through the middle of town.

There must have been rain, and plenty of it, and now all the rain that had fallen on the fells around Windermere was draining out, travelling furiously back along the Leven, which cuts a short, narrow channel given the 18 miles of lake that it drains from. After this point the river kicks north for a final, tightly curving loop before turning again and pouring out into the bay.

Minutes after entering the wood, we saw that water had begun to encroach and was filling the ground, finding hollows and drowning the bases of the oaks, ashes, hawthorns, hollies, beeches, buckthorns, hazels and yew. We were walking on a well-made path but our feet were sloshing along and I remember that we were incredulous. We walked further and the depth of water grew steadily around us. There came a point when we realised that soon the dogs would be swimming and neither of us relished the wet Wellington boot effect either, so we turned and paddled a retreat. We hadn't known it then but given certain conditions Roudsea floods with a mix of river and seawater. Given a combination of the highest spring tides, significant rainfall in the Windermere catchment and south-westerly gales in the Irish Sea, you might find yourself stranded, forced onto higher ground and waiting hours for the road to open again.

* *

I went to Roudsea again recently to have another look, to see what I remembered and to find out what it's like now. My pals Brian and Astrid were voluntary wardens. We met in the car park at the edge of the wood. There'd been a high tide and parts of the road had clearly been inundated. I mentioned my first visit and

Astrid responded by showing me film she'd taken on her phone of the river pouring like a waterfall, filling drainage channels and spreading out over the metalled road and onto the dip at the bottom of the car park, cutting off the only road out. On that day they had been stranded, but happily so. They had planned for it, and with the car to sit in and keep warm they'd watched the levels rise and inundate the woods and then waited until the levels fell again.

We circumnavigated the wood and it seemed familiar, as if I recognised the direction of the path from years ago. Now though, much of the path is boardwalk over the mossy, boggy sections. The path looped and dived back inside the wood and soon enough I began to feel disorientated.

At the side of the path deep in the wood we passed a small group of stone huts. They were either well built, or had been restored. The woods had clearly been used for coppicing to supply the local charcoal industry. In a coppice plantation the trees are regularly chopped down almost to their stumps to promote many new stems shooting up. Once the trunks reach a certain size they're cut and used as fast-burning fuel in the charcoal burner pitsteads. The spin-off activity of bark-peeling was carried out by family groups who came during the summer months and stayed for periods of time. The bark was sold to tanneries and used in leather production. As well as the traditional conical charcoal-burning pitsteads, there were also similar-shaped huts, but these were intended to be more comfortable, complete with fireplaces, a place to be dry and warm in the British summer, and as Brian said, the women needed their home comforts. In the 19th century the workforce was supplemented by the poor of the parish, widows or orphans and foundlings sent into the woods by local landowners to cut and burn bracken. The resulting lye was caustic, and one can only wonder what the process did to their lungs and eyes.

Many of the animals, plants and insects found in the woods here are rare elsewhere. The tiny hazel dormouse, so rarely seen, is on its north-west limit in Europe here on the fringe of Morecambe Bay. Much of the coppicing activity that still takes place at Roudsea is to support the survival of this tiny creature. I'd seen one only once, on a walk at Allen Banks in Northumberland, a favourite weekend walk during my years in Newcastle; it helped to restore me after a week of working in inner city housing estates. There'd been a family – grandparents with two small children – crouching down and looking intently into the grass at the side of the higher path above the River Allen. The man invited me to come and have a look, and there, apparently oblivious to our giant, looming faces, a dormouse was working, weaving a nest out of short staples of dry grass in the heart of a clump of long green grass. When eventually it was satisfied, it curled up and fell asleep whilst we uttered soft oohs and ahhs.

There are otters and brown hares at Roudsea. Nightjars, hawfinches, sparrowhawks, woodcocks and marsh tits too. And a moth – the rosy marsh moth – long presumed extinct but found again here in 2005. On warm spring evenings its caterpillar feeds on the leaves and buds of bog myrtle. There's a particular spider too, the raft spider, capable of killing small fish and even fully grown dragonflies.

We walked close to a pylon where a nest platform for ospreys had been built. A pair had come in two years ago and tried to breed, but the nest platform was destroyed by high winds. Undefeated, they came back this year and raised two chicks – another successful pair for Cumbria.

We seemed to walk for hours on paths or boardwalk that led over and under and around the mosses, or raised hills, rocky ground and even passing caves. I'd begun to think I'd never see the bay again. Maybe Roudsea did work a kind of magic; I wouldn't be surprised. Overhead the sky was turning orange. These clear winter afternoons, when the days are short, produce such dramatic skies towards the end of daylight. At last we came to the edge of the bay and walked out of the woods onto a grassy, sodden margin. The tide hadn't been high enough to encroach into the wood, but it had clearly covered the marshy fringes. With the tide recently ebbed, the muddy edges of the bay were wet, glistening, speaking. We stood in the middle of a small bay looking south over the sands. The light was changing, deepening to pinks and reds. There the viaduct dissecting the Greenodd estuary was a dark line against the rosy sands and their illuminated, sky-reflecting channels.

I walked onto the mud, just a couple of feet or so, and that was far enough; the stuff was so wet and soft that my boots began to sink almost immediately. I followed a line of indecipherable tracks leading out onto the mud, but they came to an abrupt end; something here had taken flight. Shelducks gathered on the side of a channel. Across at Greenodd, traffic moved at speed along the road where it passes beside the coast. The number of times I'd driven along there, never thinking that just across the inlet was this beautiful, secluded bay, and this wood, and dormice making nests, spiders eating fish and ospreys making new lives.

⋆⋆

There's something unsettling about finding horses, and large ones especially, in a field with a footpath running through it. We stood at the gate eyeing each other up, the two horses, me and Steve. They seemed OK. They liked having their noses rubbed and I guess they were hoping for the bribe of a carrot or a mint – neither of which we had. Fifty yards on and there was the stile leading out of their field: the Cumbria Coastal Way following the flood embankment to Foulshaw. After my visit to Foulshaw on that perishing cold spring day, I wanted to walk along the embankment to join the dots, but now it looked as if we might not be able to get through. I looked around to see if anyone had spotted us and might offer some reassuring words. There was a solitary cottage behind us. We'd walked past it and seen a pair of ginger cats sitting on top of a wheelie bin, and a car on the gravel, but no other sign of life except a bright pink sheet on the washing line.

We figured the horses must be alright so eventually I opened the gate and moved through, closely followed by the dog. As soon as we did, the horses started tossing their heads and moved towards me. In her usual state of panic with the unfamiliar, Milly ran circles round my legs. The horses put their heads down to inspect the dog; she really wasn't doing us any favours. With the horses four times my size, I thought better of it and retreated again, beaten.

We retraced our steps back along the footpath towards the unnatural and unrelenting mechanical drone of the pumping station that drains the farmland of Ulpha. Beyond it the main drain ran straight out to the bay, carrying away water from the fields. To the seaward side of it, pools of seawater remained, probably left behind by a recent high tide that had inundated the fields. A heron took off from the channel and further along a cormorant on the steep banking held its wings out to dry.

At Ulpha, at the very edge of the bay, there's a wood on a craggy knoll named, appropriately enough, Crag Wood. We went into the wood through a gate, hoping for another route towards the banking. A sign on the gate read, 'The Woodland Trust – Welcome'. We felt better already.

It had been a strange autumn. Plenty of rain, cloud layers blocking out the sun most days, and no frost to speak of yet. The trees seemed to have taken on the mood by turning as one in a sudden shift from green to uniform brown. There'd been none of the fiery and dramatic deep reds or golden yellows of most years.

We walked into the wood through leaf litter up to our ankles, noticing the same phenomenon here of the leaves yet to fall being a uniform dull brown. There was a sense of the primordial about the place though; it felt untouched, unmanaged. Fallen trees remained toppled, leaning, caught on neighbouring branches or crashed and broken on mossy rocky outcrops. I guessed that this showed the unbroken strength of the wind here as well as the age of the trees. But more too, the importance of new environments made from old, of fungal spores radiating out along dying timber, colonising and creating life. Beetled bark undermined, peeling, dying, and giving nutrients for insects that depend on dead or decaying wood for part of their life cycle. I'd read that around 13% of animals and plants in the UK depend on deadwood habitats. There were the usual broad-leafed woodland trees: mostly oak, some beeches, birches of course, as well as holly, with gorse at the fenced limits.

A few minutes' walk through the wood and we came to the edge of the bay and a wooden bench. What a place to sit and be, with the wood to your back and the bay in front. A curlew lifted from the edge of a channel close by and took to the air, warbling its alarm call. Further out on the bay another curlew lifted and together they moved laconically towards a more distant channel. Small cliffs gave way to the sands, and rocks were jumbled below at the sea-edge. Oak trees grew out from the cliff top, holding their limbs over the sands, and the bases of their trunks grew more horizontally, tight to the ground as if ballast for the weight carried above. Their roots were partially grounded and partly clinging, exposed, to the crag itself. Many of the branches that cantilevered out over the sands had become broken, or rather had been torn off; life here was hard.

The muddy sediments of the bay were wet right up to the edge of the land, though the water in the channel seemed to move in two directions at once. We watched a while, taking bets on whether the tide ebbed or flowed and then, with the result unproven, we wandered on. At the far side of the wood we found a straightforward way over a stile and through to the footpath. We'd come at just the right time; much earlier and the way ahead would have been blocked by the sea. Large puddles remained in the grass where the tide had been. I negotiated my way around a wide pool then I was on open land again. There was the field and the horses, their backs turned towards us and the bay.

Steve began throwing sticks for the dog into the edge of the tide. She was doing that collie crouch, eyes fixed on the stick, and within minutes of chasing she was filthy. We walked on towards Foulshaw, the land here a mixture of fields for cattle and stands of scrubby willow. Walking on the top of the bank, our view was expansive. In the middle distance the characteristic pale surface of Foulshaw Moss, beyond it the great stone bluff of Whitbarrow. Although it was only just after two-thirty, the sun was lowering over the bay, colouring the newly exposed mudflats yellows, oranges and greys. Huge shafts and bands of sunlight fell onto the retreating sea and were reflected back. Milly wanted to be in the water but it was too far out now; we threw the stick onto the very rim of the bay and into puddles of water to wash her clean again. She bounded on up to where the path passed between stands of gorse.

Water wicked down into the oozing mudflats. A heron took flight from its position underneath Crag Wood and drifted out towards the railway viaduct where it crosses the bay from Arnside to Grange. Light seeped underneath the supporting pillars. I heard a train in the distance, and as we turned for home, it came rattling over the bay, sunlight from the far side illuminating each window as it passed above the sea.

**

On our way back through the trees I saw something through a screen of low-growing holly, something lying in the field next to the boundary of the wood. I scrambled out of the wood, making my way down a twisting path into the field. An oystercatcher took off from rocks below. Working its pied wings furiously, it jinked away along the coast.

I walked towards the remains of what seemed a large fishing boat. Each of its two decks were split wide open, its diaghram a mass of broken and split timbers. It was as if it'd been attacked by some mythical-sized beast and had its innards pulled out. The boat must have been in the field for years, quietly imploding in on itself. I wondered how and by what means it had come here. It was close enough to the bay, only a matter of yards, to have been floated up on a high tide and it was evidently old, though I doubted it'd been abandoned in the times when the bay was still navigable.

Milly raced on and began sniffing the ground around the timbers. I passed a raft of detritus spread on the ground, oil drums rusted to the colour of red earth, concrete blocks, sheets of corrugated iron and wooden fences subsumed by the grass. Trees, especially the birches, had started to colonise the field and willow was growing in the damper earth. Briar stems wrapped themselves around the piles of stuff and there were rosehips, lush and red and out of place.

Someone had placed a rough wooden structure up against the boat as if for a gangway, though I couldn't imagine the timbers were able to take any kind of weight at all. The gunwales had all but disappeared so the hull timbers protruded like ribs from a carcass. The wheel-house had gone completely, assuming there had been one, and the remains of what looked like a broad mast poked up, sawn off and level with the fore-deck. In the middle of its shattered deck, a section of keel had burst through like a knife in the boat's back. There was something wretched about a boat this size, lying broken-backed, abandoned and burdened by land. It seemed an ignoble end.

The hull was a grimy Indian red colour and a dividing band of white paint described the boat's circumference. Above this it had been painted a bright turquoise-blue, though this only increased the sense of poignancy I felt; that someone had chosen this optimistic shade for their boat... and now look. Further round still and the planking of its hull was springing open as though the strain of resting at this angle was too much to maintain. There were areas of the hull where the paint was even more distressed; it had become fractal, cracked and the edges of it were lifting. It was as if a bloom of green lichen was spreading, colonising, not fading and disappearing. Timbers from the boat lay scattered on the ground, wood that had been steamed and pressed into shape, manipulated with skill and attention to detail. I thought again of an animal carcass, the remains left behind once the scavengers had had their fill.

But there was something else too. As I came round to face what remained of the deck, I saw again what I'd failed to fully take in. Growing from the heart of the boat, the place where great rents and clefts were filled with broken timbers, a grove of trees had taken root. Six birches, that great coloniser, their slender white trunks growing tall and straight up out of the wreckage. It brought to mind the Spanish galleon found preserved, or petrified, in the swamped forest of Gabriel García Márquez's great novel, *One Hundred Years of Solitude*. This was an image that had stayed with me since I'd read the book years ago, the galleon found during an expedition to connect the community and their village to the wider world. For the central character, José Arcadio

Buendía, the galleon was a symbol of the sea being close by, yet it was never found. The party returned to their community, defeated. Here at the edge of Morecambe Bay was as great a mystery: the finding of a boat in a field with a wood growing from out of its heart.

In the tops of the birch trees and held at their very tips, a small constellation of remaining green and gold leaves, pinned into place and just on the turn. Later I thought of the world turning slowly on cold, clear winter nights underneath the Plough, Orion's Belt and the Pole Star, witnessing and guiding the boat to its ultimate end; the long, slow journey back into the earth.

Ten

In Search of Lynx, Elk and Wolves

IMAGINE THE BAY not being there. Imagine standing on a hilltop looking out at a great plain that extends as far to the south as the eye can see (though maybe you don't know it as south). The plains are rimmed with hills and forests receding into the distances. From this vantage point you look over the grasslands and in the distance pick out a single, gigantic elk, or herds of horses grazing the rich grasslands, unaware of predators watching, wolves, or man, staking out the kill. Dust trails as the chase begins. Imagine the decisions: which herd to follow, which spear to use, which animal to single out for the kill.

* *

I went looking for a cave, or rather, more than one. I'd read about Kirkhead Cave and Kents Bank Cavern – the news had broken just a year or two ago that human bones excavated from Kents Bank Cavern, near the village of the same name, were officially the oldest human remains found in northern Britain.[17] Two of the caves are set in a steep-sided, complex limestone valley and there's a third on a sheer escarpment amongst layers of receding cliff edges in thick woodland. Underfoot the terrain is difficult;

although there's the approximation of pathways, much of the ground is clogged with mossy boulders and rocks, fallen trees and brambles. Without a guide and good light you'd be hard-pressed to find them.

It was mid-November. The time of year when, if the sky is clear, the light is at its most dramatic and intense, providing short but memorable sunsets. I'd gone to Kents Bank and, walking up the hill from the coast, the view over the bay opened before me like a map. The tide was out, the channels were there, filled with water and snaking away towards the Irish Sea.

Finding my way onto the land wasn't exactly straightforward, and after walking up and down a suburban road trying to find a way onto the fell, I saw a man clipping his hedge. As he crossed to my side of the road and stood back to admire his handiwork, I asked if he could tell me the way.

'It's literally just there,' he said, pointing at a driveway past garages and a farm gate. 'I don't know if it's a public right of way but that's the gate.'

'I've come to have a look at the caves,' I said.

'Oh yes, I know. They're not easy to find though,' and he set about describing the most direct route.

'Stick to the north side of the fell. Follow the line of the valley, keep north. I've been in them myself – took my lad down when he was a teenager. I heard someone's covered them over with brushwood to put people off going inside.'

'Why would they do that? Are they dangerous?' I asked.

'Not what you'd call dangerous, but you need to take care scrambling down into them. The worst part is getting at them. Don't leave it too late,' he said, looking at the sky, 'there's only about an hour of daylight left.'

I walked onto the headland. On the summit there's a Victorian folly, Kirkhead Tower, built of local limestone – it's a brilliant viewpoint. The light was monotone, the bay rendered cold and hard, all was steely grey. I climbed up the few remaining steps of the folly and, as I did, the sun burst out from behind a bank of grey cloud and sunlight flooded into the bay from the south. The low-lying fields at the edge of the bay turned a bright, intense green as if released from anonymity. Adjacent, I spied the roll-back form of Humphrey Head, its wooded, gentler face towards me.

I moved down into the woods. Thin trackways had been worn along the ground – badgers' foraging routes. A young deer crashed out of a thicket and bounded up the cliff edge and away in seconds. The winter song of a blackbird drifted up from the direction of the village and a distance away I saw sheets of bright light reflected from suburban windows. I pushed past holly trees, the bright red of their berries the only colour in the wood. The whippy stems of young beech trees filled the spaces between older trees, and there were ancient dark yews, of that dense, light-denying, midwinter, midnight green. I found a fence and a stile but decided not to cross; the land beyond it fell steeply away. I'd hoped for easier terrain.

The wood began to darken. Through the vertical spaces I saw low sunlight illuminating the water in the bay, but in the wood there was already a sense of the light dimming down. I found a couple of orange ropes suspended from thickened branches overhead. They made me think of that great freedom of my generation's childhood, and the way this was so often diminished these days. At least my own kids had the woods down on the Green to spend hours in. Now my eldest, Callum, had rediscovered the outdoors and was spending his evenings up on the fell and in the woods, building bonfires and taking seriously the business of being out there. Top of his Christmas list this year? New fell boots. That's my boy.

The caves eluded me. Using the map, it seemed they should

be here, just here, but there was no sign of them. I felt useless, unsystematic, and defeated. I remembered what the man up on the road had said about the caves being purposely hidden, and as I walked the terraced cliffs backwards and forwards, and finding nothing and retracing my steps, I decided to give up. I needed more specific information, and maybe a guide. As I came out of the woods onto the open fell, a deer shot ahead of me and back into the woods again. A train rattled around the coast, though I couldn't see it. Out on the open fell-top again the light grew more intense. As I moved towards the tower and the gate back to the road, my elongated shadow made a path for me to walk upon.

★★

'I wouldn't be surprised if this whole area is covered in caves, given the limestone. We'll never know though.'

I had my guide. I knew that my friend Brain Hardwick had been acquainted with the caves and that he'd been inside them, but what I hadn't realised until we spoke recently was that he'd been involved in digging and surveying two of them during archaeological investigations in the early 1990s.

'Must be 15 years since I was here. I was slimmer then, of course. I'd be down on my belly slithering along and into some tiny spaces. I pulled a huge length of rope out from inside the back of one of them – badgers must have dragged it in for nest material.'

We planned to meet on the first Sunday of the New Year. 'If we don't go then, the weather's going to be against us for the rest of the week,' Brian had said. His wife, my friend Astrid, came too. We set off down the hillside and into the scrubby undergrowth and came to the stile that I'd stopped in front of last time. Brian stepped over it and disappeared over the lip of the steep slope, Astrid and me following in his wake.

Kirkhead Cave was recessed underneath an overhang of limestone cliffs. Once we reached the terrace of earth outside it, we scrambled over boulders and leaning trees and down into the lip of the entrance. The cave was last officially investigated in the early '80s, though it wasn't until seven years later that the floor was covered with a protective fabric membrane and gravel, to protect further layers of undisturbed deposits.

The place had clearly been used by local youths, and fair enough. But the place was a mess: ubiquitous plastic bottles slung into corners, redundant tea-light candles on rock shelves, even black bin bags filled up with rubbish but left in a heap at the entrance. Astrid and I despaired. In a far corner someone had left a tidily rolled duvet and a folded towel, as if they might reappear at any moment.

'Come and look at this,' Brian said. We wandered to where he stood, looking up at the ceiling of the cave.

'When it was discovered, it was nothing like this size. It'd filled over the centuries with earth and deposits.' He was looking intently at a low section of roof. 'All this carbuncle-looking stuff is calcite flow. And here...' – he pointed at a sheer roundel of rock wrapped in the yellow-white of calcite – 'there's been a big stalactite broken off from here, probably by Victorian enthusiasts. They have a lot to answer for.'

One of those Victorian enthusiasts, John Bolton, published an account of his explorations of Kirkhead Cave in 1864.[18]

> *I have been acquainted with it for about ten years; my first visit being in 1853... We found the height of the cave at its mouth to be three feet; consequently admittance could only be gained by crawling in on hands and knees. Beyond the mouth, the height of the roof varied from eighteen feet, at the part nearest the entrance, to twelve feet; the length of the cave we found to be forty feet, and its width twenty feet;*

the area consisting of one irregularly oviform chamber.

During excavations of the top few inches of soil deposits he described finding a Roman coin: indication, or proof, as Bolton thought, that the cave had remained undisturbed for the past 1,800 years.

He described his two labourers digging out an area that measured seven feet in depth over an area of around 50 feet. They found the bones of birds and badgers, fox, goat, pig, wildcat and the tusk of a boar. There were axes and animal bone fragments thought to have been inscribed with simple patterns. Then:

At the depth of four feet, a portion of the right parietal bone of a human skull was thrown out. Continuing the excavation to a depth of seven feet, we obtained another human bone, which proved to be the second lumbar vertebra, and the radius and ulna of a young human subject.

Towards the base of their excavation they found significant quantities of blackened sticks of wood, potentially showing that they'd reached the oldest layers, and evidence from the late Interglacial period and of potential funerary rights:

There was also found a rude bone implement resembling a knife, a piece of carpal bone of goat two inches long, having a round hole through it, as though it had been suspended as an amulet; together with several fragments of pottery rudely burnt, similar in composition to ancient British cinerary [funerary] urns.

They also found the hollowed-out bone of a pig, similar to finds from other locations that had been made into primitive musical instruments. I'd heard one played at the Kilmartin

Museum in Scotland, its spectral sound whistling down the aeons.

Much later on in the 1960s further work was carried out in the cavern and finds included further remains of a stalagmite floor, flints and other implements and evidence of human activity that dated back to the Late Upper Palaeolithic and Mesolithic, as well as intermittent activity in the Late Neolithic and Bronze Ages associated with the interment of human remains. An antler boss of Megaloceros, an extinct giant deer akin to the Irish Elk, was recovered in 1969 and yielded a radiocarbon date of over 10,500 years of age.

⋆⋆

We left the cavern and walked north and into a valley I hadn't known existed; I'd not walked this far in the dimming light of my first visit. The valley floor sloped back up towards the fell in a tumble of thickly mossed boulders. The place had the feel of orc country. Brian led us scrambling up an increasingly steep incline to a shallow terrace and there, unseen from below, was Kents Bank Cavern.

'This is the one I helped to dig out. When we were here it didn't have a name, it was just a hole in the ground.' Two caverns burrowed into the earth, converging and connected by a hole too small for us to scrabble through. We switched on head-torches. Astrid moved into the deeper, right hand tunnel and Brian and I entered the left. A large, bulbous bloom of calcite covered the rock and a white, more liquid-looking sheet flowed through its midst.

'That's the more recent formation,' Brian said, and then, 'I'm afraid someone's using this as a public loo. Look – badger latrine.'

I scrambled out again and as Astrid came out of the deeper channel I went in. A chunk of timber was propped between two rock faces.

'I probably put that in,' Brian said.

I clambered down into the deepest chamber, and in the bottom of the cave the beam from the torch became diffracted, filling the small space with a globe of misted opalescence, a halo of artificial light following my gaze. Deep in the bottom of the furthest space another channel no more than 18 inches high led away into the dark. Brian was behind me now and, pointing at the aperture, said, 'That's where I found the rope. I ended up with a cut finger too – when I looked at what I'd grabbed, it turned out to be a flint. No wonder it was sharp.' I thought of Brian's hand being the first to take hold of that flint cutting edge in over 9,000 years. I'd have gladly suffered a cut for that.

'The dig was run by Chris Salisbury,' he said. 'He lived back on the road up there. We used to climb over his garden fence to get down here. I was brought in to do some of the donkey work, the uncomfortable stuff. Chris was the brains. You could have done with talking to him, but not much chance of that now.'

'Why?' I asked.

'He got cancer, poor bloke. The thing is, most of the stuff he got out hasn't been seen since he died. No one knows what happened to it all; it seemed to disappear into thin air.'

* *

Looking over the small, deep valley we were almost at eye level with the third cave, known as Whitton's Cave. Evidence of at least three further human interments had been discovered in this cave together with a shard of primitive decorated pottery beside one of the skulls, thought to have been Bronze Age. But once again, looking up information in the records, the whereabouts of the finds are mysteriously not known.

The three of us scrambled down the valley side and back up again. Astrid dived straight inside. This cave had a level entrance accessed by a recess in the rock, like a lean-to natural porch, a home from home. I struggled with my rucksack straps and faffed about trying to find my head-torch. I was about to follow when Astrid came out again and looked me straight in the eye.

'I went in so far, and I'm not going any further.'

I thought she'd found a body. 'What's the matter?' I asked.

'There are spiders. Huge spiders.'

I went into the cave to check for myself. There were a series of chambers with linking arches, each one successively lower than the last. The third arch implied a tight-fitting squeeze into the furthest chamber, and I didn't want to be denied access.

'I can't see any spiders. Well, not big ones anyway,' I said, peering at the roof of the chamber where a few small arachnids crawled over the stone.

Astrid came in behind me.

'Not there – there!' She pointed to the narrowing entrance into the furthest chamber. Giant spiders covered the archway.

My family will tell you, I've always struggled with spiders. On a sliding scale there are some I don't mind, and even some I quite like – the wispy, long-legged Huntsman spiders with the tiny round bodies that come in the summer. But big spiders and I don't mix. At summer's end, when the females come into the house looking for a place to hole up, I go into the rooms at the back of the house with a wary eye. If there's an intruder, I'll catch it in a plastic cup and take it outside, and if no one's looking I encourage them to go and live next door by shaking the contents of the cup into the neighbour's garden. Squashing up close to these cave-dwellers was just not going to happen.

Whitton's Cave had a warmth to it. Situated on that side of the valley wall, I guessed it must catch the afternoon sun, and the cave air was filled with small flies. I'm sure it was a really good place to live – if you were a spider. Astrid and I peered into the furthest chamber and then Brian arrived, expressing indignation at our attitude to the occupants. He moved in, took close-up photographs and came out to show us. That was close enough for me.

We left the cave and its inhabitants behind and clambered up the valley, a place that seemed ancient, hidden and untouched. We ducked underneath fallen trees and climbed over boulders before emerging into the light on the top of the fell again, and five minutes later we stepped back onto the suburban road with its clipped hedges and orderly winter gardens.

⋆⋆

Before we left that day Brian suggested I get in touch with another member of the Chris Salisbury team – Dave Coward. 'He kept detailed notebooks. He'd be able to tell you more about the place.'

Later that day, I re-read the report detailing the age of the remaining human bones. The remains had found a secure home in the Dock Museum in Barrow. On the website I found Dave's name associated with the finds, so I got in touch. A day later, we talked on the phone.

'Chris was very much a "gentleman archaeologist",' he told me, 'a maverick, an old style academic. He'd not have been out of place 50 or 100 years ago – I mean in the way he related the stories of discoveries from the past. He was following in the footsteps of John Bolton, the local Victorian "earth scientist", known for writing *Geological Fragments of Furness*.'[19]

I told him I was familiar with Bolton's reports of Kirkhead.

'Bolton was working at a time, I forget the exact year, when his discoveries and the thirst for knowledge amongst the public were so strong, they had to find a big enough hall to fit everyone in. Up to 1,500 at a time went to listen to his talks. His samples are in the British Museum and I believe are even now a major part of the reference collection for that period.

'But as for Chris, well, there'll never be digs again like the ones that Chris ran. We had no risk assessments or hazardous substances statements; nothing like that. I remember questioning Chris about this, because it's something I have to deal with at work. He said, 'I'm not joking when I tell people they can sue me; I've got nowt so you'll get nowt.

'After Chris became ill, I asked his wife, Anne, if we should stop – but she said to keep going, it would be a distraction. I used to visit Chris in hospital. I'd find him sitting up in bed drawing some of the finds from Kents Bank. He kept them in a box in his hospital locker. After he died, they weren't seen again.'

Thankfully some of Chris Salisbury's work on Kents Bank Cavern had found a secure home at the Dock Museum and were eventually tested by a team from John Moore's University in Liverpool. Amongst these were pieces of skulls from at least three individuals and also a femur, a radius, an ulna from the forearm and two humeri from the upper arm. Humans were here in Lakeland from 12,000–13,000 years ago, evidenced from the finding and dating of stone tools. But the femur found at Kents Bank was subsequently radiocarbon-dated to 9,100 to 10,380 years ago, placing the individual to the early Mesolithic period, making the Kents Bank Cavern bones the earliest dated human remains found to date this far north in Britain.

It's likely that the cave had been used as a place of interment. The bodies could have been laid out here unburied, and might have been subject to excarnation, where the flesh is removed and then the bones left at different specific important sites across the landscape. This would have left the remains vulnerable to removal by carnivores or indeed to trampling by animals. Again, this is one of the earliest and most northerly pieces of evidence of Mesolithic mortuary practices in the British Isles.

But what of the other finds? As well as the human remains,

the jawbone of a lynx was unearthed at Kirkhead Cave. Lynx became extinct in the UK about 1,800 years ago, but I found it exciting to think of these beautiful, lithe creatures roaming over my home landscape. There were the bones of a 7,000-year-old dolphin that had most likely been dragged up to the cave by an animal. And a short distance away, in fact just an arrow's flight from Kirkhead Tower, is Humphrey Head, the place where the last wolf in England was reputedly killed. And although this idea might be fanciful (how on earth would anyone know where the very last wolf died anyway?), there's evidence of wolves right here on the doorstep. Elk had been roaming northern Britain over 1,300 years ago and their bones were found in the cavern. They had been gnawed by a large predator. The most likely candidate? Wolf.

Eleven

Testing the Sands

On that day, after weeks of mild weather and hardly any rain, the Kent looked nothing like a river. How different its character from the clear, voluble stretches of it that tumbled and splashed down from the Kentmere hills, passing through the valley and woods and close to our house in Kendal. It was the same element and there the comparison stopped. The river was wide and glossy, more a shallow, sky-reflecting lake than a flowing body of water as it arced towards the south on a lazy journey, passing Silverdale, Carnforth and Morecambe. The tractor pulled up at the river's edge and we climbed down, then rolled our trouser legs high. Barry handed us each a sturdy wooden stick and each of us carried a bundle of brobs (or marker sticks), hoisting them onto our shoulders. I left my brand new camera in the trailer. Andy was made of sterner stuff and took his along.

The men analysed the condition of the river, and from their talk it was as if it was a living being to them.

'A wind from the north-west like today pushes the river slow and wide,' Cedric said, 'but a big wind and heavy rain makes it cut a direct path, and it'll move at speed – much deeper too.'

I've spent significant periods of time walking alone in the mountains. I've been lost and disorientated in mists, but this was something new, this sense of flat space where distances are difficult. We were about four miles out on the bay, halfway between Humphrey Head and White Creek at Arnside. I wasn't alone and of course I couldn't have been in safer hands, but there was something about this landscape and our remoteness from land that made me uneasy; it was a new sensation for me.

Cedric climbed back in the tractor, shouting as he did, 'I don't like this bloody tractor one bit. I don't trust it either.' And he drove away, leaving the three of us behind in the middle of that expanse of sand.

★★

I'd planned an early morning rendezvous with Cedric Robinson, the nation's only Queen's Guide to the Sands. We'd talked on the phone the evening before, and I found him still giddy, fizzing even, from having lunch that day with the Queen and Princess Anne on their official visit to South Cumbria.

'We were asked to go into the dining room and find our place names. Well, I found two! One next to Princess Anne, and the other next to the Queen. So I went to find a steward and I told him, and do you know what he said? He said, "Well, Cedric, it looks as if you will just have to choose for yourself who you sit next to."

'So,' he said, 'I chose Princess Anne. And she was absolutely lovely, and do you know she was a very good conversationalist.'

I told him he's a great one for hobnobbing with royalty.

Cedric's status in the Morecambe Bay area is legendary. He's been Queen's Guide to the Sands for more than 50 years and his knowledge of the tides and the rivers that shift their route through the sands – sometimes by a mile or more in one night – is unsurpassed. In his 80s, he has the physique of a man half his age; he's tall, tanned, (or weathered), big-chested and upright, a man in his element and comfortable in his own skin.

Cedric was going out to test the sands and the state of the River Kent ahead of two crossings later that week, one for a group of horse and carriage enthusiasts and then a weekend cross-bay walk. Any crossing depends on the weather, but more than that on the amount of rain that's fallen on higher ground. In a bad summer like the previous four or five, more walks are cancelled than take place.

We planned to meet at Humphrey Head, a limestone outcrop projecting into the bay from the low-lying fields of the Cartmel peninsula.

* *

Early the next morning I drove to Humphrey Head. Rounding a bend in the narrow lane a leveret was sitting mid-road, warming itself in the sun. It moved off distractedly, lolloping ahead with black-tipped, pale-edged ears acting like vertical radar, switching direction continuously. Articulating its powerful hind legs slowly, it began to jig from side to side like a wind-up toy – all eyes, ears and legs – and then disappeared under the shelter of a hawthorn hedge.

In the small car park underneath the cliff the quality of silence was immediate; a rare moment. The engine ticked as it cooled and a wren *chit-chitt*ed in amongst young hazel and hawthorn on the cliff face, otherwise all was utterly still. Poking above the marsh grass to the south-west, the turbines of Walney Windfarm were illuminated by the low sun, glimmering like a row of distant candles. A woodpecker called from the scrub close by. I could see Hoad Monument, the hilltop lighthouse folly at Ulverston, standing out from the shadowed hills behind it. I'd learned the word 'glas' from the Welsh poet Gillian Clarke, a potent word for the particular blue-green of hills. There it was, defining the white monument, the glas of Kirby Moor.

Standing beside Humphrey Head there's an illusion that the surface of the bay slopes upwards, rising higher than the land around it. It's a trick of the light and the land and the vast, flat distances. Towards Flookburgh I noticed a puff of smoke drifting upwards from way out on the edge of the marsh grass, and at first I thought it a cloud of birds rising up into the air. Seconds later, a mechanical hum came travelling over the distance, and the silhouettes of three tractors appeared. They were putting out faint diesel plumes into the air as they ticked along the grassy horizon like miniscule automata.

Time passed and Cedric hadn't arrived. I thought it unlikely that he'd be late. The minutes ticked on, turning into a quarter then half an hour, and two more tractors followed the others out onto the bay. I phoned Cedric's home; his wife, Olive, answered.

'Hi, Olive, I've arranged to meet Cedric at Humphrey Head at nine o'clock. I'm wondering if I've missed him.'

'Oh no, don't worry. He only left home 10 minutes ago.'

I sat down to wait, taking in the cliff face and its windblown swathe of yew, whitebeam, hazel and rowan.

In the distance a train racketed along the line into Furness. A car arrived. Two women climbed out and a young girl wearing pink taffeta and wellies. They moved slightly away from the cliff then stood looking up at it through binoculars. A crow passed over and, as I watched its shadow, I heard the burr and hum of a distant tractor, the sound passing in and out of hearing as it drew closer.

Forty-five minutes after our agreed meeting time, a spindly, aged tractor bounced into view with Cedric at the driver's seat, pulling an eccentric-looking jalopy, part Child Catcher's wagon from *Chitty Chitty Bang Bang* and part ancient farm wagon. It was all patched plywood and weather-blasted Perspex.

'Sorry we're late,' Cedric said. 'I had to borrow a tractor. Mine's gone in for a service.'

I told him not to worry, and climbed up into the trailer where his two helpers, Andy and Barry, sat on a row of back-to-back benches. The floor was strewn with laurel branches. We set off. Within seconds, Cedric stopped the tractor and was down on the ground checking the best route across a deep gully, and then again a second time. With a large shovel he dug out a smoother path, but we still held on tight as the tractor and trailer plunged into and out of each gully, bucking and bouncing before reaching smoother ground.

I looked up above the cliffs of Humphrey Head and saw there, swivelling on the air, a peregrine. It folded its wings, dropped a distance, then rose up again on the morning thermals. So that's what the family had come to see.

'They're doing well here,' Barry said. 'The young have fledged; we saw them a few weeks back.'

Travelling out onto the bay there was a new sightline of the headland. The rocks rose out of the sand, then continued upwards in folded waves of limestone. With a row of fenceposts like tiny spines along the ridge, it reminded me of a sea creature emerging out of, or plunging underneath, the bay.

The tide had been out for hours but a layer of water remained like a second skin over the sands and in places there were deeper pockets and shallow lakes of standing water. The trailer rode into and out of channels like a ship ploughing the ocean. In the lee of the headland, a solitary egret stood white and motionless, like marble. Further out, the surface of the sands changed and became smoother. We made a wide curving entrance to the bay.

Another shift in terrain and the trailer began to bounce up and down. We held on to the seats again, passing small cliff edges that had formed and collapsed again as big waves had washed against them. Then came an area where the sands had formed into mounds that were like the moraine of a retreating glacier.

'I call this the Somme,' Barry said. 'These mounds come from the weight of the tide passing over areas of softer or more compressed sand.'

I saw cockles heaped up within the mounds, their edges semi-exposed like treasure.

'The past few cold winters have finished them off. They're mostly dead, no good to anyone,' Barry said. The beds had been closed since the cockling disaster in 2004; the echoes keep travelling.

The surface changed again and we moved on over a skin of perfectly becalmed water that reflected the sky, so that as I looked down we drove over intense cobalt-blue and cumulus cloud streets. The tyre marks fractured the surface, distances elided and light spooled on the sinuous ground like heavy white blossom. I remarked on the way the surface changes from one area to the next and Andy said, 'It does, all the time; every day sometimes. Over at Arnside the sands are as smooth as glass-paper just now.'

The guides have used laurel branches for centuries for marking the safe crossing routes over the sands. Their thick leaves stay on the branches even when submerged daily by the tide. They're known as 'brobs'. They're there in Turner's paintings of the bay. We drove past brobs that had been put in position on previous visits. From a distance they resembled people who had somehow been left behind, lone figures adrift on the empty bay.

'That one – the one all on its own – we call it the "man brob",' Barry said, as if he'd read my mind.

* *

The three of us were standing at the edge of the river in the middle of the bay. Further downstream, where Cedric had stopped the tractor, I watched as he wandered, looking, stopping, and setting new brobs. I began to wonder what time the tide was due in and how long we would be out there.

A helicopter droned in from the east and swung in closer to have a look at us. Resisting the temptation to wave, I wondered what the people inside thought of us, standing at the edge of a river in the middle of the bay, miles away from land.

Barry and I paddled into the river. Andy was already midway across and taking photographs. The waters changed gradually from shallow and clear to grainy and textured, coming up above our knees. I stopped to hoist my trousers higher, then, taking slow steps in the cloudy water, I let out a cry of surprise.

'What's up?' Barry asked.

'A fluke!' I said. 'I stood on a fluke.'

The squirming sensation had caught me unawares; I'd trodden on a fluke, the flat-fish commonly found in Morecambe Bay.

'Last time that happened I was just a kid. My pal's dad took us out to tread for them, but I didn't fancy the idea of standing on any.'

The traditional way of catching flukes is to feel for them with the feet, then reach in and under the water to pick out the fish before it has a chance to wriggle away.

A moment later there was the distinctive bounce of a fish on my leg, but I was OK; I was orientated. I wondered what, if anything, the fish thought about it.

Twice in the crossing Barry stopped to test the flow rate with his low-tech device: he simply watched the speed of the water pushing past his stick. We waded further and after some minutes came to the other side, throwing the brobs down in a heap. Andy arrived alongside us, and working together, the two of them began setting the brobs into place. Andy pushed and wound his stick far down into the mud.

'Ready?' he said, then pulled out the stick as Barry pushed a brob down into the loosened sand, securing it as the mud and water closed the hole again.

Over our heads a small group of young gulls came winging in. They flew as if intoxicated, directionless, mob-handed, adolescent, their cries argumentative against the complete silence of the bay. I watched as they flittered down onto the sands and joined a long line of birds that had settled in to feed on the surface a distance away. Looking through the binoculars, I saw that there were hundreds, maybe thousands, of oystercatchers. I'd never seen so many in one place before. Stitched in amongst the oystercatchers were countless gulls. Out there on the bay, it was the undisputed kingdom of birds.

I'd visited Cedric at home on a freezing February afternoon as a grey and pink mackerel-backed sky spread over the whole bay. Sitting in front of the banked-up fire at Guide's Cottage, my cheeks growing hotter by the minute, he'd told me his name for oystercatchers.

'I call them "sea-pie". They're wonderful birds. I've watched them riding the bow wave when the bore comes in. Fifty years ago the bore was a very different kind of a beast. You could hear it coming a mile off, with a three-foot standing wave at the front and sea-pie skimming the top of it. What a sight that was.'

Cedric drove back towards us. He climbed down and left the engine sputtering. 'It's changed its course again... do you think?' he called across the river. Barry and Andy shouted agreement. 'Significantly. Moved about half a mile, I think.'

'What's that sticking out of the water? Is it a bird?' Cedric asked.

'I'll go and have a look,' Andy said, and began walking down-river towards a dark object at the river's edge. It didn't fly away.

'It's a brob alright,' he shouted back to us. He pulled it out and reset it on the river-bank.

'We set them in pairs,' Barry said as we watched. 'They're like channel markers for shipping, so it's clear where to cross the river.'

★★

A dark rain cloud came close, pushed along by the westerly wind. We wondered if it would release its load onto us as we felt the first large splots of rain. But it skimmed past and headed Morecambe way. As it moved, it grew darker and minutes later bands of heavy rain fell from it. I could imagine families on the imported sand beaches at Morecambe stuffing towels into bags and rushing off to the cafés until it passed.

Cedric came wading back across the river towards us, singing at the top of his voice. If Cedric formed a religion, I might be tempted to follow.

With the four of us together again, Cedric marked the time and the men made calculations about the tide times for the coming days. It's crucial, of course, to get it right. That's part of the reason the men were there: not just to test the riverbed for quicksand and mark safe crossings away from it, but also to mark the time and to know when to be back on dry land again. It was eleven-thirty. In another four hours the place we stood would be submerged beneath 10 metres of water.

★★

Barry's a blow-in of 11 years from Manchester and he wanted to know the names of the mountains that framed the view at the back of the bay. I named them for him.

'That's Fairfield, Red Screes, then the gap of Kirkstone Pass, then Caudale Moor and the hills of Kentmere; Froswick, Ill Bell and Yolk.'

Cedric walked past us and said, 'I've never heard of any of them.'

I wondered if he was having a joke. But maybe not; after all, the bay is his territory – he's on intimate terms with 120 square miles of tidal estuary.

I said, 'I like the fact that you can see the source of this river, the mountains of Kentmere, just there.'

Cedric pointed to a wide bend in the river a distance away. 'See the way light sparkles on that stretch of water? That shows the river's moving much faster down there.' And I did, but a question formed, of how Cedric could be replaced. You can't archive this stuff, or create websites for it. You can't communicate about this place by email.

* *

Andy and Barry continued setting brobs. Cedric and I began to wade back across the river towards the idling tractor and we talked about the continuation of the over-sands route that crosses between Flookburgh and Ulverston. It involves the crossing of two more rivers, the Leven and the Crake. We'd stopped to talk mid-river, and I found that my unease and that sense of unfamiliarity had all evaporated, moved away like the dark cloud, and all the while the river pushed on, pressing gently against our legs.

Oystercatchers slung past us in arrowhead formation. Nodding in their direction of travel, Cedric said, 'We'll head over to that bank for our coffee.' I couldn't see a bank, but I knew that the bay was anything but flat. More pictures came to me from the winter's afternoon by Cedric's fire.

'There are banks and gullies out there; great holes big enough to swallow a tractor, a double-decker bus even. We'd go night fishing with tractors, depending on the state of the tides, fishing for shrimps. One night we were driving along in the moonlight and all of a sudden matey on my right disappeared, tractor and all. He'd gone straight into a massive gully. A 'melgrave', we call these big holes. Anyway, he managed to climb out alright but the tractor was another story. We never saw it again.'

He described too how, after yet another episode of the unnervingly heavy rains that we've had over these past few years, he'd gone out to assess the state of the river. He said he'd been left almost without speech, and that's something.

'The river had cut a new channel overnight, *six miles* away from its previous course.' He described the river that day as 'like a roaring sea'.

⋆ ⋆

The tang of coffee filled the trailer and Cedric held court, giving us tale after tale. His is a rich, deep seam of powerful memories, and he offered them to us with generosity, for our entertainment and I thought, for himself, for the vigour of remembering.

'I'd gone out shrimping at night with our Jean. She was about 12 then, I think. It was as clear as a bell when we set off; the stars were all out, and the moon, and you could see the lights of all the villages around the bay. It was a good night for navigation. We were miles out, and busy with the fishing, then when we came to go home I looked up and the fog had rolled in. We couldn't see a thing. Well, that night I navigated home by listening for the sound of the river. That did the trick. Anyway, we lived to tell the tale.'

⋆ ⋆

Immense cumulus clouds streets had continued to form around the edge of the bay. They built height over the land, leaving the sky above the bay a clear and potent blue. I've seen this so often, the sky clearing, as if putting itself in order, ready for the approach of the tide. As we set off for Humphrey Head again, the headland appeared like a wave swelling out of the sands, and in the distance Peel Castle shimmered above translucent air. In the heat haze the hills of Furness were breaking up into segments that moved and danced.

On the journey back to land we passed close to the 'man brob', and as we drove past Barry said, 'A few weeks ago we came out and I could see something odd about the shape of it. As we got closer, it took off. It was a peregrine. He'd been sitting there in splendid isolation until we came along.'

We were back, rolling over the saltmarsh and bumping into and out of the two gullies. Underneath the limestone cliffs, I climbed down. We said our farewells and the tractor disappeared up the lane. I listened as the hum of its engine faded off into the countryside until all was quiet again. I had a sense that those hours out on the bay would stay with me for a long time, glimmering like the river as it moved by degrees further and further into the distance.

See page 2 for a note on Cedric Robinson's death in 2021 and on the appointment of his successor, Michael Wilson (now King's Guide to the Sands).

Twelve

Humphrey Head

THE TRACTOR HUM FADED AWAY into the distance and I walked towards the west over the marsh. In places a slew of seawater remained from the high tides, making ingress up towards the flood banking.

Turning to look back at the limestone headland, I saw how it rolled out of the land in the long rise of a whale-back, anchored into place by the saltmarsh and the geometry of fields. I'd looked at the shape of it, seen from above on my computer at home, projecting out into the bay like an arrowhead. And I'd seen how its profile resembled the tiny, sharpened microliths from the late Mesolithic period that I'd handled in Lancaster Museum, tools that sat comfortably between my thumb and forefinger, and that felt as if they'd been fashioned to fit my fingers exactly. They were coloured of the bay's landscape, the place they were discovered in: amber and limestone-white, grey, and black as obsidian.

I walked out towards the bay again, following a linear channel at the base of the cliffs. Barry had told me about a spring that emanated from a fissure in the rock. There was nothing to mark it now other than the merest trickle of rust-coloured water. He'd said that the waters were once thought to have health-giving properties, that there had been a house under the cliffs and that people came from miles away to take the cure. 'Drink it now though, and all you'll get is a gippy tum.' Then Cedric said that a mate had once taken a drink just to see. 'He said it tasted like iron, it was disgusting.' I found it, nothing there more than a slight drip of water and broken bricks, greasy red-earth stain on the stones.

Further along a memorial had been carved into the rock:

Beware how you these Rocks ascend
Here WILLIAM PEDDER met his end
August 22nd 1857 Aged 10 years

I looked up and wondered about the odds of climbing the scrubby cliff face to the top. There's a huge cave and rock arch high up on the cliff, though it can't be seen from below. I would have liked to get at it, though I didn't much fancy the climb alone; I'd need to research a route. A slight wind moved through the rowans and yew and there again, hanging in air, was the peregrine. Its eye worked the ground with tiny movements of the head. It hung so still, it was as if the air it balanced upon was a solid entity. Eventually, rising higher still, the bird swivelled and banked and disappeared from sight over the headland.

Where the creek widened and the limestone became too steep to scramble over, I took off my trainers and socks and paddled. Under the clear water, warm mud oozed up between my toes. I was suddenly aware of being alone and there was, I noticed, the slightest feeling of vulnerability; yet I hadn't even reached the open sands.

Where the headland petered out to a jumble of large boulders, I sat down amongst sea-pinks, sea holly, daisies and grasses. Splashes of chrome yellow lichen covered the rocks and rabbit droppings littered the fissures between them. Looking out, had I drawn a straight line over the sands, the next landfall would have been Heysham's two nuclear power stations, box-like and brooding.

The creek widened further, moving out into the sands in a series of bends. I tried to locate the place where we'd set brobs by the river, or where Cedriç had stopped the tractor for coffee, but from that low viewpoint it was impossible to be sure.

A heat haze appeared near Morecambe and through it the facades of hotels and blocks of flats broke up, becoming fractal, and traffic fizzed along. Just a couple of weeks previously I'd spent some time on the island of Mull and whilst out on open water towards the Cairns of Coll, we'd seen dark abstract shapes forming and dissolving along the horizon. At first I'd thought it must be the island of Barra, the closest island in the chain of the Outer Hebrides, but the forms began to dissolve. They continued changing, shape-shifting, one minute making mountain ranges, the next becoming an elongated egg then stretching out and becoming thin to the horizon again. I knew this was some trick of the light that I wasn't able to explain, maybe something to do with the way the optic nerve receives information, but I thought perhaps the islands were being modulated, as if there on the line between sea and sky they were magnified and shrunk again as lightwaves bended over the surface of the ocean.

A wide margin of saltmarsh filled the distance between me and the northern horizon. With the perpetual nature of change on the bay and the channels shifting their positions, the saltmarsh now grows where previously all had been sand. At Grange the derelict open air swimming pool was surrounded by an amphitheatre of grass. I'd been there a few times as a kid. Maybe we took the train, though I doubt it as that would've involved a bit of walking and my parents didn't do that, not if they could help it. I had two clear memories of it though: one of the mind- and body-numbing cold water of the pool; and the other of looking out from the top tier of stone seats at windsurfers scudding past and waves lapping at the stone base of the building. Now the area around the pool was transformed, turned into pampas lands of ochre and green. Every few years there's talk of restoring the pool again, making it into one of those places ubiquitously called 'attractions'. If it ever happens, I just hope they make the water warmer. Whatever, it wouldn't be quite the same looking out onto a sea of grass.

There was a good perspective of the rising slope of limestone of Humphrey Head and the sudden drop of the cliffs on its western side. A swathe of dark green hawthorns spread towards the deciduous woodland of the eastern slopes. I began to walk up the rise of land, passing through a kissing gate and following the spine of a fence onto meadowland. Two hawthorn trees marked the graphic nature of the prevailing wind; their trunks were bent completely horizontal and smaller branches rose up from them vertically in knotty clusters.

Even a small degree of elevation opened the view to a panoramic of the bay. A solitary white egret on the edge of a channel seemed the only living thing in that wide sweep of space. With the binoculars I located the flow and twist of the River Kent. The tracks of the tractor were still visible and I was able to chart our journey out to the river and from there it was easy to locate the brobs we'd set into place just a couple of hours ago. The breeze was growing stronger and I knew that a rising wind often signalled the returning tide. I followed the silvered kink of the river to where it dissolved to a shining thread and widened again into the forked tributaries of a delta. On the Irish Sea two ships ploughed the deep channel like counter-weights. Out there, sparkling and illuminated by the bright sun, was a moving edge of water. I looked at my watch; there were three hours to full high tide.

★★

I thought of the legend about the last wolf in England having been killed here on Humphrey Head. It's a dramatic enough setting; I'd buy the idea of it at least. There are other claims to the location too. I'd read of a last wolf legend in Devon, but they'll tell you anything in the south. Below the headland there's an interpretive sign about the last wolf story. As a child I had my fair share of literary indoctrination that identified wolves as slavering, mad, granny-eating beasts. On a rare family holiday in the Welsh village of Beddgelert, I'd learned the story of the medieval Prince Llewelyn the Great's faithful Irish Wolfhound, Gellert. Whilst hunting, the prince realised that his dog was missing, and on his return home he found the animal smeared with blood. As the dog went to greet its master, the prince saw that his infant son's crib was empty. Thinking the dog had killed his child, the prince plunged his sword into the dog. A child's cry was heard, and Llewelyn found his son, covered by a blanket. Next to the child lay the body of a great wolf. Faithful Gellert had protected his master's infant son from the wolf, and the prince realised his terrible mistake.

Although he wasn't the first monarch to call for the desecration of wolves in the United Kingdom, King Edward I of England commissioned the killing of every single wolf in 1281 as an economic expedient, not only to protect sheep and other livestock, but because of the number of deer being killed by wolves in royal hunting reserves. The extinction of wolves became a reality. Given the rank antipathy of some landowners and gamekeepers even now to endangered species, for instance towards hen harriers that breed on moorland, it's hard to imagine how landowners or sheep farmers would take to the re-introduction of wolves. But this has been called for by top zoologists over recent years. Bring back top predators, control deer numbers and native forests will regenerate themselves.

Jim Crumley's book *The Last Wolf* deals with the debunking of the myth of wolves as savage, senseless serial killers:

> *And the truth about the wolf – one truth about the wolf – is that it can be like a bridge in all our lives, a bridge where enlightenment may cross, a bridge to a place where we don't make all the rules and where our species is not always in charge. And if some people are disadvantaged by our willingness to allow the proximity of wolves back into our lives, people like sheep farmers and hunters, then that is simply part of the price that we pay for the privilege of a closer walk with natural forces, part of the debt that we owe for all that we have taken out of nature for far too long. We cannot rationalise every decision we ever make as a species on the basis of whether or not it will be good for the economy; sometimes the greater good of planet Earth must come first, and the wolf, as the master-manipulator of northern hemisphere ecosystems, is an agent for that greater good.*[20]

I like too his description of wolves as painters of mountains. The reintroduction of wolves to the Yellowstone National Park in the USA has seen huge regrowth of native species – of plants, trees and animals. 'In the long wolfless decades they [the deer] forgot how to behave like deer. Then the wolves came back and they remembered.' The land was able to regenerate, became painted with colour once again.

★★

Mares' tails and high cirrus merged with vapour trails and there was the glint of an aeroplane sewing a line of white in the intense blue. Above Morecambe an immense cumulus cloud grew a hammer head that moved slowly towards the upper atmosphere. It continued until, miles high, it began to reform and resembled the plume from a volcano. Dark on the far eastern horizon, the familiar anvil-topped summit of Ingleborough. Closer in, the 'man brob', isolated and uncanny, like one of Antony Gormley's human sculptures set out in the middle of the bay. A lone curlew made a curved circumnavigation of the headland; the wading birds were beginning to move back to the bay from their summer breeding grounds in the hills.

The tide came on, filling first over by Morecambe, and slowly travelling north to join the River Kent. The water filled deep troughs and poured over sandbanks. In places it isolated areas and came back for them some time further into its flow. I watched the place we'd walked through the river's shallow banks become submerged. I thought of it picking up the river water, of the river's fight for direction and the mix and flux of it all going on beneath the surface while above it simply sparkled and shone.

As I reached the trig point at the highest elevation on the headland, the water had formed a parabolic arc around the headland. Then a movement by the cliff edge caught my eye: the peregrine, working the wind with all the dynamic force of its feathered engineering. The wind had risen blustery and the bird fought it, used it, dipping and diving in currents and eddies of air.

The breadth and complication of the bay was slowly being unified by that travelling sheet of water. Sandbanks were surrounded and subsumed and the sea moved on north, flowing under the piers of the viaduct at Arnside. Adjacent to Ulverston, Chapel Island was gradually surrounded. I wanted to stay on that airy viewpoint watching the tide shift and backfill until it was complete, but I had to move on; there was a birthday to prepare for and a cake to bake. The next day Fergus was to be 13. A faint line of oystercatchers flowed out from the land towards the vast distances of the bay, and I turned for home.

Thirteen

A Tale of Two Hills

IT WAS A BRIGHT COLD DAY close to the very end of December. Steve and I had come to walk off the effects of Christmas, along with some Ulverston friends and their children. We met up on Birkrigg and set off walking the common's eastern rim, where the view leads back over the fields and the Croftlands estate towards the town, and beyond that the frozen grey-white backdrop of Coniston's ice-bound mountains. We dropped down by the bridlepath between drystone walls into the village of Bardsea and on to the edge of the bay. It was mid-afternoon and the light was changing all the time from flat grey to sudden intense bursts that turned the sands red. We walked along the shore below Sea Wood, then clambered up the raw earth bank that was knotted with exposed tree roots, some as long as branches, and on through the trees to return to the lower part of the common.

At the stone circle we stopped for a while. Looking back to where we had walked, I saw the village church's white steeple pointing skywards from a circle of trees and beyond that, offshore from the great woods and stony foreshore by Conishead Priory, was Chapel Island. At either end of it sandbanks were slowly being inundated by the gathering tide. Further still, the wooded slopes of the Cartmel peninsula pointed into the bay, the piece of land that separates the two main river channels of the Leven and the Kent.

I grew up thinking the stone circle was a folly, something built on a whim. After all many of the town's gardens and footpaths feature pieces or even whole walls of limestone plundered from Birkrigg Common or another of Cumbria's outcrops of this characteristic weather and water-sculpted stone. But the circle is absolutely authentic. It is one of only 30 concentric stone circles in the UK, the best example of course being Stonehenge. This circle on Birkrigg is the only one of its kind in Cumbria, and forms part of a more complex picture of Bronze Age life in this part of the bay. Elsewhere on the common a number of tumuli, or ancient burial mounds, have also been found. The inner circle of 10 stones is self-evident, the stones reaching a height of just less than one metre, but the outer circle, made up of 15 stones and with a diameter of some 24 metres, is largely hidden in the bracken or overgrown by turf.

Excavations in 1921 unearthed five cremation sites, three in pits, one on some cobbles and one covered by an upside-down urn. There were a few small but significant stone implements found too, identified with possible ceremonial uses: a pestle, a palate and a piece of red ochre.

Being there beside Birkrigg's stone circle brought to mind another of Cumbria's remarkable megaliths, Swinside, off the road from Broughton to Millom. I'd planned a visit there many years ago, and decided to walk the route in reverse so that the stone circle came near the end of the walk and was a treat in store. Walking with my dog, as we came closer to Swinside Farm I came across a series of unequivocal hand-painted signs: 'No Dogs Allowed'. But to walk back would have meant a two- or three-hour return journey without having reached the stone circle, and more importantly, with no food or drink left in my sack. I walked on, keeping the dog firmly on the lead. Skirting the flank of Swinside Fell, the circle came into view. I crossed the farmyard, almost holding my breath in case the irate farmer came out to shout at me and my dog. Looking over the wall the circle was made up of over 50 stones, some upright and some that had fallen, almost like a jawful of teeth in varying stages of decay. I love its other name, Sunkenkirk, brought about because of the local legend that whenever anyone attempted to build a church on the site, by morning the Devil had pulled the stones down and created the circle out of the spoils. There are well-documented examples of ancient sites being hijacked for new places of worship.

I liked too the idea that massive changes affected the celebration of rituals, with the building of open rings superseding gloomy chambers or covered tombs. As Aubrey Burl, the megalith specialist and archaeologist, wrote, the civilisations building stone circles were making a massive shift 'from darkness to light, from the dead to the living, from the grave to the sky'.[21]

Walking down the farm lane, the way ahead was blocked by a tractor, and the farmer was digging out a drainage ditch, chucking out spadefuls of great clods of wet black earth. I was prepared for the assault but decided to get in first with a cheery greeting. Almost an hour later I was still there, unable to get away from a man who clearly didn't get an awful lot of company, and who, having explained the number of out-of-control dogs who'd attacked his sheep in the past, eventually waved me off with a friendly goodbye. He even gave the dog a stroke too.

* *

On Birkrigg Common, someone said, 'Hold hands inside the circle. See if we can feel the cosmic energy,' and as we did so a drone came flying past, remote-controlled, black, insect-like, and anything but cosmic. As we set off again, the machine returned to its owner, descending into last summer's desiccated bracken.

As we reached the top of the common I saw a line of geese travelling in from the north, distantly, silently. They were side-on, but flying in characteristic V-formation as they appeared as two slightly interleaving lines. They covered the miles between the town and the common in slow-winged minutes, then circumnavigated the hill, moving east towards the bay and dropping out of sight. As they did, a scatter of rooks lifted from the roof of Sea Wood's umbered branches. They dispersed, moving out over the bay. The wedge-topped summit of Ingleborough, the white Portland stone of the Ashton Memorial tower above Lancaster, the power stations at Heysham and the ancient stone tombs on Heysham Head passed underneath the tilt of their wings, then the birds coalesced, banking back over the wood and they melded, pouring over the fields and coming to ground a field away. Wings ruffled momentarily and the birds strutted the furrows, then settled and begin to forage in the earth. A train approached from the Cartmel peninsula, crossing the viaduct over the Leven channel then disappearing again into the contours of the land. A few minutes later it emerged again and rattled towards the town.

From the summit of the common, the soft hills and fields of Furness rolled out from beneath my feet, falling towards the hamlet of Sunbrick in its shelter-belt of trees, the geometry of farmland, the villages of Stainton and Gleaston sunk in the limestone dales and a castle ruined before it was completed. My gaze moved to Roose, where the shipbuilding town of Barrow began to form, then along the coast to the winking flare of yellow burning off from the inshore gas terminal at Roosecote, to Rampside and the house with 12 chimneys. To Foulney, that thinnest and most insubstantial of islands, which on that day was just a painted line upon a sea of silver-grey. To Roa Island and across the narrow channel to Piel Island's teardrop, the smallest of landfalls at the very edge of the bay.

Once only accessible by crossing the sands, Furness is even now accessed by only one major road and away from that a fusion of deep, lark-filled lanes, bordered in season by May blossom, hawthorn and blackthorn, each lane affording a new perspective on the land or the bay. As remote as you'd like it to be.

To the west great banks and bands of blue-grey cloud deceived the eye. In their midst was what might have been another land; it could almost have been just another cloud-country, glimpsed between the layers. But then we saw the summit of Snaefell and the neighbouring hills on the Isle of Man, just a line of sea visible between it and our coast. I remembered that on a really clear day it's possible to see six kingdoms from here: England, the mountains of Snowdonia in Wales, the Dumfriesshire hills of Scotland, the Isle of Man, Ireland, and the Kingdom of Piel Island.

To the south-west, where Walney Island marks the end of Furness and of the bay itself, the Irish Sea seemed to flow into the sky. Both sea and sky were rosy, that particular cerise and orange glow of a late winter afternoon. Out of this light the windfarm emerged, a trigonometry of stationary blades. It is vast, marching across the sea as if pegging the land into place, or preventing the glacial moraine from sliding underneath the sea. Beyond it, across the Duddon estuary, Black Combe marks the swing to the north of the Cumbrian coast.

I looked across at the smooth, rounded summit of Kirby Moor, below it the dip in the hills where the town of Ulverston sits, backed by Flan Hill, its bracken-sloped, ancient, zig-zag, hawthorn-hedged fields marking the route north-west towards Gawthwaite Moor, the Coniston hills, to Broughton-in-Furness and the secret hills and valleys of Woodland. It was to Woodland that my father and I travelled every Sunday morning during my final year at home. As Dad sat drinking coffee and catching up with the market gardener and his family, I explored the fells and lanes by horseback.

Underneath Flan the stream of Gill Banks runs through open, footpathed fields where we caught minnows as kids in calm pools above a small waterfall. For some reason we called them snotty bullies. We took them home, keeping them in washing up bowls until they became horribly inert. Then Dad took us to release them back into the beck where they began, ever so slowly, to nose their way upstream. Below the field a metal gate and the stream dives into the deep, wooded valley of Gill Banks, one of those formative places of childhood where we sallied for hours, twining each other up on rope swings and unravelling in a headlong, sickening but compulsive spiral, over and over again.

Hoad Hill unwinds itself above the houses. On the summit the lighthouse and below, like a conduit from the monument to the sea, the line of Ulverston Canal. At its end the sealed lock gates that speak of another era when, from up here on the common, there would have been the sight of ships in full sail veering out to sea on the edge of the wind.

The lighthouse calls to the far north, to the ice-locked latitudes where the town's celebrated son, Sir John Barrow, sent men to seek a navigable route through the Canadian Arctic Archipelago and on to the Pacific Ocean and the trading nations of Asia, to worlds of plenty. The monument calls to John Ross and to James Clark. It calls to John Franklin and the men of *HMS Erebus* and *HMS Terror*, sealed up in ice in 1845, succumbing slowly to scurvy, hypothermia, starvation and lead poisoning.

But the lighthouse also calls to a less well-known figure, and points to warmer latitudes. Whilst stationed in the Cape Colonies, now South Africa, Sir John married Anna Maria Truter, a remarkable botanical illustrator, who compiled the first known catalogue of Cape flowers.

By the time I had turned a full circle, the glow by Walney had begun to fade, but closer in, a single sunbeam broke from the clouds. Falling vertically it pierced the sea near Aldingham. A few minutes more and the beam had shifted over the water and slowly begun to dissolve. We dropped north-west from the trig point down the broad, steep pathway that runs down towards the road and the cars, and then moved on to our friends' house to sit by the fire, eat more Christmas food and talk.

★★

Another winter's day and I walked out of the centre of Ulverston by its northern artery, turning right into Chittery Lane, the final road of houses where there is a view over the town then on towards the bay and the south. The road becomes a lane and at its end divides and goes off in two directions. Straight ahead there was a walled garden, a single, delightfully situated allotment where an older man looked at the ground, a plastic sheet in one hand. He looked down at sprout tops, at leeks and at kale.

I turned left, walking along a walled lane and at its side was a stand of dark fir and sepia-toned larch. The tips of the tallest larches, wind-bitten, bent over at 90° from the trunks. On the other side of the lane, the fields dropped steeply away over the rooftops and there, just a stone's throw away, was the roof of the house my father built in Whinfield Road. Out to the west the green, remembered hills of Kirby Moor that I looked out onto from my bedroom window.

The town clock chimed the three-quarter hour, the sound fading as the lane turned right and passed beside another larger and darker wood, where bikers had made a trackway with planks for obstacles amid the deep, sienna-coloured leaf-litter. Then there was the Victorian kissing gate, sagging on its hinges. A sign had been attached to it: 'When the flag is flying on the top of Hoad Hill, the Monument is open,' then, 'Welcome to Hoad Hill. Drinking, Graffiti and Skateboarding are Strictly Prohibited.'

I wondered if I'd be rumbled, if sniffer dogs would find the flask of hot chocolate buried deep inside my rucksack.

Through the gate I walked out onto the open fell, the path winding ahead of me and leading to the top of the hill and there, glowing a soft white against the early morning sky, was the lighthouse. It appears like something from a book or a story, but I don't question its presence. Having grown up here, of course I don't. Its intrinsic eccentricity fits, this lighthouse without a light a mile away from the sea. As I grew older I began to understand that there was something particularly eccentric about the town and some of its indigenous characters, as if all had been part of some grand plan, or put together by a novelist with a flair for individuality and place. Ulverston was like nowhere else on earth. Even the sheep on Hoad Hill seemed to belong here, with the yellow patch over their backs, the ruddle mark from the tup to prove he's done his job, but also pink marks on either side, a blue mark on the top of the shoulders, black faces and curly horns.

Overhead the clouds were red-edged, pewter-coloured, but blue patches were beginning to form. The cold tingled on my cheeks. A slight mist had begun to form over the town, occluding certain buildings and streets. The path wound on past slabs of rock and low-growing whin. For me the hill has a sense of friendliness about it. Can a landscape be friendly? Perhaps it's just the memories I have, that Hoad Hill, being so close to our childhood home, became somewhere that made sense and that offered no threat. Indeed it became somewhere that offered stability. Somewhere that grown-ups, or more specifically, our grown-ups, never ventured. It didn't matter what was happening at home – and that could often be difficult – here was an escape route and a place that was reassuringly constant. So yes, I think a landscape can be friendly. Perhaps even supportive; or at the very least, sustaining. And crucially, it was the place I can ascribe to my formative relationships with wild places.

Where the fell rose up and the path contoured along its side, there was a stand of three or four larch trees growing amidst a group of rocks. When I'd walked here as an adult, I thought of it like a Zen garden, an arrangement made by nature that has a feel of perfection about it. I looked back along the way I'd walked to see the path as a series of curves and bends and the landscape's drystone walls parcelling out the land into workable pieces. The path climbed and then the view north towards the Coniston hills began to open out. The monument called me on. There was a group of people up there, a couple of adults and a handful of children. The children were doing exactly what all children do up there; they were running around the wall at the base of the monument, round and round and round. Cresting the final pull up to the summit, the sound from the road below came in, the A590, the main artery connecting Furness to the rest of the world. And it was the road that led back to where my mother believed happiness was to be found.

Over the rush of traffic I detected a different kind of noise and a train rattled over the viaduct, making landfall at Tridley Point, the sound of it echoing through the open spaces as it rolled in towards the town. Tridley, the place by the edge of the bay that a school friend and her mother had taken me one day for a picnic, and me leaving home feeling slightly guilty after my mother said she didn't understand what the attraction was. 'It isn't the real sea after all, or the real sand.'

Then I was standing at the top of the hill with the monument beside me. And what a view! These days there are useful interpretive boards to pinpoint and name all the visible landmarks. Across the low-lying fields the bay, the Leven channel, the viaduct, then anvil-topped Ingleborough, Darwen Moor, Winter Hill and on south past Blackpool Tower and into Wales to Cyrn Y Brain and Moel Famau, closer in again, Birkrigg Common. Then further to

Moel Siabod, Carnedd Llewelyn, Eldir Moel, Faw Eilis. The hills of Wales were not visible that day, but their names made up a litany of another place.

I approve of this naming business. When we first came to Ulverston and began making trips into the Lakes by car – only ever exploring by road – it was the names of things I wanted. I wanted to know the names of the mountains I saw, of the passes, of every hamlet and hill. I wanted to have those names at my fingertips so that each time we drove past I could look up and say to myself, 'Fairfield' or 'Helm Crag' or 'Stickle Pike, Harrison Stickle and Side Pike,' and for no other reason than I felt a need to know. There is, of course, a musicality to this, a kind of notation that is special to the area and is, in its own particular way, as significant as the Welsh language is to the Welsh. I felt, in acquiring the names of features in the landscape, that I was also acquiring the language of the area, the language of the Lakes. Give me a map even now and I'll soon be sidetracked away from the job of searching for a route to instead finding the place names and their musicality.

The children and the two mums set off down the hill again. One of the girls said, 'I think the cows have gone now,' and looking in the same direction I saw at the far end of the hill one of the hardy cows that sometimes gather up here, just because they can.

From beside the monument it's possible to trace underneath a pointed finger the streets and places familiar to me, and in a small town like this most places are familiar. At the base of the hill, and where the footpath leads towards the town, the huge piece of pilfered limestone resembling a rearing elephant, then the gardens and grounds of Ford Park House, part of my old secondary school and now a community centre and playing fields. Into town and the route to and from Dale Street Girls Junior School, which appeared to have been run by a triumvirate of ageing, wing-tip bespectacled women. Possessed of the kind of twisted personality as if from a book by Roald Dahl, amongst them was fearsome Sweaty Betty, and though I can't recall her real name, I remember well the vicious lunchtime attacks, stabbing some unfortunate girl in the back with a set of painted red, sharpened finger nails, for failure to eat up every scrap of liver and onions. These days you could take a case against her to the European Court of Human Rights.

Into Fountain Street and past the Walker Brothers' old shop, which became a magnet for Angela and me. Pony mad, we'd call in after school to stand at the rough wooden counter as the 'younger' Mr Walker fetched pieces of saddlery and bridles, halters and horse bits we'd asked to see – how he entertained us I'll never know. He'd disappear through the door into the warehouse and as the sound of his feet faded away, we'd look at the 1950s posters for Cherry Blossom shoe polish and heel replacements – all glamourous women and new heels and Brylcreemed men smoking pipes. Mr Walker might be gone for ages searching for whatever it was, but back he'd come. We'd hold and examine and talk forever about whatever the thing was, and then we'd leave again.

I followed the line of the street to its confluence with Soutergate, the narrow road to the north, past Angela's house, then further up the hill and, there, the roof of our house in the new estate on Whinfield Road.

Somewhere, I have a postcard. It's a picture of me, riding my bike through the middle of town, past the market cross where, unknown to me, someone had taken my photograph. I'm riding my three-speed bicycle with the wicker basket attached to the handlebars that had been concocted from many parts by Eric Bibby, the bike man. For all I know I might've been sent round the world on that bike; or at least to Manchester, or Cleethorpes, or Luton. Come to think of it, it was the best bike I had for years.

I looked up at the clouds moving against the solid, unmoving tower, conspiring to bring feelings of dizziness, of disconnection, of not having one's feet solidly planted on the earth. Yet it was this place, this town, where I firmly planted my feet and which sustained me once my parents had moved back to Merseyside. I came back to Ulverston in my early 20s after two attempts at college, neither of which worked out. Whenever I could, I walked in the mountains or went down to the bay, to the places where I felt I intrinsically belonged.

And then I stayed put, working at a variety of jobs and also working out my place in the world, and what it was that I needed, and where I needed to be. Although I didn't yet know it, I was asking myself the question, as the poet Mary Oliver asked, 'Tell me, what is it you plan to do with your one wild and precious life?'

★★

Before I left the monument I saw another sign: 'Enjoy England, Place of Interest'. I was glad that someone pointed this out – I suppose it keeps them in a job. I turned and retraced my footsteps down the path and this time, instead of looping around the far end of the walk, I cut off left where the winding path runs down between two narrow drystone walls and comes out again on Little Hoad. Here there had once been a secret garden and a cottage screened from sight by tall conifers. We'd sneaked in one day, my brother Andy and other friends, and then we'd hidden behind dense shrubs as an old man began to call out, 'I know you're there.'

I looked through the screen of trees: the cottage had gone, burned down. Inside the garden a solitary robin sang a wintery lament. I walked down the final field, coming out by a terrace of tall Victorian houses and on, walking past the lamppost at the edge of a small wood where I'd once thought I might find Narnia, and where we'd thrown ropes over the crossbar at the top of the lamp to turn it into a swing. Below there, the churchyard of St Mary's, and a memory of walking there one Sunday morning before anyone else was awake, of picking snowdrops to take home for Mum. Making her breakfast in bed and taking everything up to her on a tray.

I wonder though, about how it was that my parents seemed so unable to make a connection with the landscape. How could you come here to live and not learn to love it? How is it possible to ignore the landscape, or to disdain it, or to feel no sense of its emotional pull? But I have Harold Wilson to thank for making it all possible for me, for the happenstance of my father's redundancy and finding work in Barrow. Now I've boys of my own and when one of them says (and they frequently do), 'Don't we live in the best place?' I feel a sense of self-righteous and wholly smug satisfaction.

Fourteen

Aldingham and Baycliff – A Ghost in the Grass

THE SOUTH WAS UNDER WATER. So much you could imagine the whole country beginning to tip, ever so slowly, underneath the weight of it all. In Cumbria we had such a dark beginning to the year: rain, grey skies, and even more rain. The land was sodden and our boots had a permanent tideline of mud that I gave up trying to clean off.

But like rare pearls amongst all the gloom were days when the sun burst through to illuminate everything, lifting the spirits. I'd planned a trip to the far side of the bay. A friend had reminded me about the small church in the hamlet of Aldingham. She'd told me that the church, the churchyard and the foreshore all had a particular quality, an energy to it, and that you could feel it. So I set off to re-visit for the first time in years.

The coast road from Ulverston follows the level terrain out of the town and into open farming country. It passes close to the Buddhist Centre at Conishead Priory, an imposing, if architecturally challenged building. The site has a significant history that pre-dates the current house. In the 12th century a hospital was founded on the site, followed by an Augustinian priory. After the dissolution of the monasteries a procession of landowners took control of the estate and there were disputes, deaths, land claims and counter-claims over the centuries until the present building was begun in the early 1800s.

The house took 20 years to build; its owner, Thomas Braddyl was bankrupted by disastrous speculation in the Durham coalfields. A veritable parade of buyers and losers followed until a Scottish syndicate developed it into a hotel. It had its own railway that connected to the main Furness Line at Ulverston, the line of it passes through the woods to the east of the priory. A long straight path leads down to the bay, passing over the disused track and a low bridge, a place where ferns populate the dank walls and water seeps and drips onto the ground. I'd walked here hundreds of times with friends and our dogs. Through the dark of the wood, the chiaroscuro trunks of hornbeam and beech and oak; beech mast crunching brittle underfoot. In winter, the spread of snowdrops, like light falling. Through the Victorian gate and out to the shore at the edge of the bay: Chapel Island, not much more than a stone's throw away across the sands and the channels.

At some point in the 1960s the building was turned into a convalescent home, ironically, for miners from the Durham coalfields. My primary school teacher, with whom I had a not altogether successful relationship, was at least possessed of the saving grace of music. She played the piano and we sang and played recorders most days. As well as musical productions in school, she arranged for us to give a concert for the miners at Conishead. Beforehand she told us about the conditions the men had worked in down in the mines and the diseases that affected their lungs. I think then that we had gone into the concert room with trepidation, wondering what on earth the men would be like; memory tells me that they were just friendly, smiley and appreciative. The room was high-ceilinged and I remember that light fell from tall windows. Thinking about that room I see the palest greens and yellows, like an aura surrounding the memory; whether this was the colour of the walls or sunlight pouring down into the room from that high window, I couldn't say.

Once the convalescent home closed, Conishead was abandoned for a significant number of years. During that time legion quantities of dry-rot spread its creeping mycelium behind walls where it ran rampant, unseen. During the 1980s the priory found new owners – a group of Buddhists. These days, after years of work to rid the place of rot and to repair the building, it is now an internationally renowned centre for meditation. There's a sparkly new temple too.

* *

At Bardsea the character of the bay has changed dramatically in the time since I moved away. Saltmarsh now extends a significant distance into the bay. The roadside car park used to give an uninterrupted view over the bay, a place where retired folk came for an ice cream, a sit and a look. Not that the sands were necessarily golden though; far from it.

There was an evening in the 1980s, it was early summer. I'd brought a group of boys from the residential school I worked in. We'd come down to the bay for a picnic and an ice cream from the kiosk and a walk. And then the boys decided to find water. They wanted to swim, but the tide was far out, and I wouldn't have let them in the tide anyway. But being the kids they were, they found water in a channel near the shore and jumped down into it, stripped to their undies, throwing their clothes onto the sand at the edge and were splashing about in the shallow trough of water. The top of the channel was roughly at boy head height. When they emerged a while later, they'd become tribe. They came, advancing in a line towards me with wide-legged, joggling steps, each one plastered from head to toe in thick, drying mud. We hadn't brought towels, and I sent them back into the channel to wash off the mud. It stuck fast. As I drove them back to school that evening they waved and smiled at everyone we passed, teeth and eyes looming white from inside their dirty Morecambe Bay mud faces.

★★

At Sea Wood the road begins to rise and passes through the trees, then follows the banks and bends and emerges into the light again at Baycliff. Here the view opens out over the whole bay. On that January day the bay was at the mid-point of sea and sand. Channels formed arabesque lines and shapes, the water within them glowed quietly in the light. At the far extremity of Morecambe Bay, the meeting place of sea and sky had fused into one metallic band of silver-grey cloud. Further west bright sunlight sparked along the horizon.

The road dropped down again. The land here is green, parcelled into farms and fields marked out by drystone walls and small lanes. I turned into the lane to Aldingham and stopped the car outside St Cuthbert's Church. As soon as I turned the engine off a memory came in of the last time I was here, and I couldn't explain to myself why I hadn't made the association sooner. Some 25 years ago a former boyfriend, Howard, had died after enduring months with cancer. On the day of his funeral I'd travelled over from my home in the North-East.

Howard wasn't a religious man but during his illness he'd befriended, or been befriended by, the vicar at Aldingham, and he'd come here to sit and talk, or to sit and look out over the bay, and to think. He'd found some kind of peace in amongst the enervating days battling disease and contemplating, as I remember he did in our conversations, the end of life. The small, intimate church at Aldingham became the obvious place for the service.

A group of Howard's close friends arranged for a memorial stone to be made and letter-carved by his friend, Boris Howarth. I began to search amongst the few dozens of stones set face-up in the grass, their intimate scale perfect in the small churchyard. Some had flowers, either recent and fresh, or faded. Others, plastic or silk, had become edge-singed through the wear of time and weather.

Then there it was. The grass had begun to encroach around the face, but in the centre, carved in curving calligraphic script his name and dates: 'Howard Steel 1948–89'. Time collapsed.

Kneeling on the grass I grabbed the turf-edge and started to peel it back, pulling off the longer tendrils that were creeping across the face of the stone, and exposing again the inscription arranged around the edge:

and the only bridge is love, the only survivor, the only meaning.[22]

The grave seemed abandoned, forgotten. I wanted flowers. I wanted to restore a sense of care, of remembering.

After the funeral service, I'd walked arm in arm with a friend along the rocky foreshore, trying to make sense of this early death, of having witnessed this big man physically shrink over the months as the cancer took over. Then a friend of Howard's came running along the shore towards us and as he came closer I saw that he was running over the stones in bare feet. I wondered if the physical pain involved might have had an effect, ameliorated his sense of grief.

⋆⋆

I went into the church and walked along the aisle to the east window, a large piece of stained glass made and installed in 1964.

The figure of Christ is central, and above him the dove of peace and a hand descending from the blue firmament; God, I imagine, with accompanying angels and their unfurling banners bearing the legends 'King of Kings' and 'Lord of Lords'. But besides all the usual religious iconography, the glass-artist and designer – the wonderfully named Harcourt M Doyle – has given us the bay.

There's an oystercatcher probing the sand with his orange bill, observed by a redshank. A black-headed gull, its wings pulled back and up into the zenith of movement, lifts itself over the bow-wave of the Morecambe Bay tidal bore as it thunders back into the bay. And the work is populated by the people who live and work beside the bay. There's a farmer, shirtsleeves pushed up in the warmth of a late summer harvest, hefting armfuls of bound wheat. A fisherman holds up his catch – a Morecambe Bay fluke, what else? The fish is silver-grey against the dark blue of the fisherman's gansey. Both men look up at the Christ figure rising over the landscape; they're unsurprised by his presence. On the opposite side of the window, the low escarpment found just further along the coast, the site of an Iron Age hill-fort, and two unexpected figures. A young woman; she reminds me of the way Leonardo da Vinci painted women, golden hair falling down her back and a rope of it coiled over her shoulder. She's holding a baby in her arms, swaddled and cosy, fast asleep. She looks to the earth as her man digs the soil. At first I can't make sense of what I'm seeing and have to look again. They're wearing animal skins. It makes me smile. It's a nod to our past, to where we come from, our beginnings, the sense that we weren't the first to be here. This is the beginning of cultivation, of farming, of coming in from the nomadic hunter-gatherer way of life and settling here beside the bay. Wildflowers and grasses surround the man's feet as he makes the first cut into the earth. It's an image of hope and expectation, looking to the future but absolutely acknowledging the past. It says that there is more than one way to look at the world, even through the eyes of faith. And I like this, this idea that it doesn't matter who on earth you are, or what era you lived through or what your belief system is; we all belong. It makes sense then, that this was the place Howard chose to come to, to be in. Above all the figures in the window, great cumulus clouds move over the bay, in just the way they do.

* *

Out into the sunlight again, a mass of snowdrops were growing in the shelter of the drystone wall, their faces fully open, probably for the first time this year. Their structure was complex, layer upon layer of petals inside the flower head of that first green of the year and edged with a tiny frill of white. I picked just four stems for Howard, placing them on his stone just to the side of his name, and arranged them with their faces turning upwards towards the sun.

* *

The sea had moved further away, leaving a milky meniscus of water. A family were walking on the sands and the way the light fell it was as if they were walking suspended, just above the surface. The boy called to his Labrador and chased it around in muddy circles. A group of redshank probed the sands and, as they moved, the reflections created the illusion of their legs being doubled in length, like birds on stilts. On an impulse I took my shoes and socks off and walked around, checking to see if I could feel the energy that Fay had talked about. Nothing mystical happened, just the cold, like the shock of stepping into a chill stream. Then I remembered the man running in bare feet over the stones, and thought that this was good, that maybe I was connecting to that final collective moment for Howard.

★★

I was feeling subdued, all this stuff coming up that I hadn't foreseen, all this unanticipated analysis of an old relationship. Then more memories came piling in. Of when I'd gone to see Howard the day before he died. I'd travelled back to Cumbria to say goodbye, though I never got the chance. Howard's friends and colleagues had gathered at his house and stayed by his side so that he never had to be alone. I wanted to say goodbye, to know that I'd done this instead of shying away, of avoiding, and living with the consequences. But how do you make conversation, or be, at a time like that? I wanted to go up to his bedroom and hold his hand for a few minutes, to stroke it gently and say goodbye. But the invitation didn't come and I found that I was speechless and couldn't ask, rendered mute. Another memory: I'd heard that the man in bare feet was the one with Howard at the end, holding him tight, cuddling him like a baby, as he'd slipped away.

★★

I put my shoes on and walked down the steps onto the foreshore where a singular, sickly smell jolted me back to the present. A fox, its eye an unnatural cavity, lips curled back and teeth squatter, blunter than I'd imagined. Washed up, apparently, and lying in the tideline that, because of the very recent storms, had been pushed up tight against the steps and the churchyard wall. A female, I thought, smaller than a dog fox and no urban fox either. Her coat was in good condition, pure rusty red and underneath that a lighter, thicker, more golden layer of insulating fur. Her brush lay stretched out horizontal as if in anticipation of speed, but had become diminished through long immersion in water. My eye made this quick initial survey and then travelled back over the animal more slowly, and the find took on a darker, more gruesome tone.

All that remained of the fox's legs was one front humerus, the newly exposed joint gleaming unnaturally white. There were no signs of other leg bones, nothing. I tried to make sense of it. What was the story? Had she been washed out from a bayside den inundated by storm waters, another victim of the three successive storms that had very recently smashed into the bay? She'd have been no match for the weight, the swell and roll, the speed of it. But it seemed unlikely that the legs would have simply rotted away. Had she been dug out from her nearby lair by some irate farmer? Having whacked it over the head, dispatched it with the spade that had dug it out to the light, had he hacked off the legs in a gesture out of all proportion?

Two women were walking towards me, each sweeping a metal detector just above the cobbles. I saw them stopping close to the dead fox. One of them unpacked a fold-out spade and began to dig, immune apparently, to the stench from the rotting carcass. Now there's commitment for you.

* *

I drove back along the coast road to Baycliff. I wanted to go down to the shore to see the rock carvings made by one of Ulverston's much-loved eccentric sons. Bill Stables was a master-craftsman. He ran a workshop from a lock-up garage in the middle of town. Inside it was a mass of stacked-up timber, machines and equipment, and from here he made traditional wooden rocking horses that were beautiful, detailed – and coveted. He made roundabouts and furniture too, all from this cramped, apparently disorganised space. Someone once asked him how long it was since he'd last been to the back of his workshop. 'About 10 years,' he'd said.

I only ever saw Bill in a black shirt, open to the waist in all weathers. White, curled hair sprang out from his huge, barrel chest. He was a fit man, walked with a swagger, chest out, shoulders back. As a child I'd seen him walking through town with his cat perched up on his shoulder.

Bill was a well-known naked swimmer. He spent most of his free time at Baycliff swimming when the tide was right, and carving when it wasn't. When Bill realised that I was a swimmer he was very encouraging. He'd been getting deafer and spoke in the over-loud voice of those who can't tell how loud they are speaking. He spotted me in town one day, and in a street packed with people he bellowed across the road:

'Come with us down to Baycliff to swim! You don't need to bother with a costume! We're all pals!'

Lucky for me I'd seen the funny side.

Once Bill had swum he would stay on the beach letter-carving into a broad sheet of flat, white limestone rock by the shore. The rocks were, in the end, a memorial too; Bill didn't want a gravestone.

I couldn't think about Bill and Baycliff without remembering my friend Marie. We'd completely lost touch. She'd been a great friend for years. We'd played tennis, cycled, swum together and spent evenings in the pub with her husband, Chris, and our friends. Her family home in the small hills behind the town was a welcoming place; there was always a meal in the oven and home brew.

The nuclear reprocessing plant of Sellafield was not very far away up the Cumbrian coast, and we'd both been involved in anti-nuclear campaigning. In 1983, it was widely reported that a team of divers from Greenpeace discovered high levels of radioactivity when they tried to block the Sellafield underwater discharge pipes. Due to concerns about contaminated discharges, the Department of the Environment effectively closed stretches of beach near the plant.[23] A small group of activists began to hold local meetings, but the scale of the action grew almost overnight after the showing of a TV programme about childhood leukaemia in the village of Seascale. A public meeting was held in Barrow – the hall packed with people, just ordinary people who wanted to be involved. Pete Wilkinson of Greenpeace came to help us get organised; he was like a rock star in our midst, all black leather and London swagger. He talked to us about making and maintaining a coherent and sustainable campaign. We needed teams of people to take on particular jobs. We stood in queues to sign up for something – direct action, publicity, administration. The atmosphere in that hall crackled with the sense of the campaign becoming something tangible, of the possibility of David taking on Goliath.

But then a woman's voice broke above the hubbub.

'There's something not right here. What about those two?'

A Barrow matriarch was standing up. She was pointing at two men near the back of the hall, arms folded over their chests.

'What have they signed up for? They've signed up for nothing! What are they after?'

There was a sense of their metaphorical feathers ruffling momentarily, but they sat on, implacable, arms folded across their chests.

Other voices joined in.

'Where do you work then – Sellafield?'

'How come you've not signed up then? Who are you working for?'

'Have you come to observe then?'

And within minutes they'd gone, melted away into the evening. It was our first encounter with observers.

Later, some of the group came to suspect that phones had been bugged. Phone calls were made to make plans for protests at Barrow Docks as new shipments of spent nuclear fuel arrived from Japan, but the police would always get there first.

But Sellafield did not merely annoy a group of anti-nuclear

demonstrators in Cumbria. Since those times there have been a number of protests from both the Irish and Norwegian governments about the plant, amid wider public concern about the risks posed by radioactive contamination.

* *

Marie and I had swum together outdoors all through the summers, and without this daily immersion in natural water we'd felt unable to function. But she was made of tougher stuff and swam outdoors all year. I'd watch her stomping on layers of ice at the edge of Coniston Water to make an opening large enough to swim in, then in she'd go, hooshing and whooshing until she came out shivering but sated, a few minutes later. We'd swum together at Baycliff but I came to it with a wary eye. I couldn't understand the attraction compared to swimming in lake water. Here you could only swim at the turning point of the tide. Come too early and the tide might pull you up the bay; too late and you might be gone. You had to know how to look at the water and judge the flow and only swim at the turning point of the tide. I liked much more the meditative qualities of lake or river water, the way your skin felt afterwards, as if the water remained like a membrane over your skin. The mossy smell lingering.

I parked the car at the top of the lane that led down to the shore and as I was gathering my belongings noticed the coast road bus coming along the road. It pulled in at the stop and a woman climbed down. She was wearing a long, black coat and bright red glasses, and white hair fell down her back from inside a woolly hat. She had bags of shopping and a rucksack. She set off down the lane ahead of me, but it wasn't long before I caught her up. As I drew level with her I noticed activist badges lining the front of her coat: 'Free Gaza', 'Close Guantanamo Now!', 'Free Tibet', 'Troops Out of Afghanistan'.

I said hello.

'Oh sorry,' she said, 'I was miles away.'

We looked into each other's faces, then time collapsed for the second time that day. We hugged each other. Questions began to come, one after another.

'What are you doing here?'

'Well, what are you doing here?'

She answered my question first.

'I've lived here for years, in one of the chalets.'

'I've just been thinking about you and remembering the swimming and Bill. I wanted to see if I could find his carvings again.'

'I think they're still there – just further along the beach. I haven't been to look for ages. I love it here – it's so peaceful. What about you?' she asked. 'Have you come back?'

And I told her about coming back to Cumbria to raise the kids, about Steve, his work, my work, the boys. We swapped stories and major events as we walked on down the lane towards the shore, talking and stopping, setting off again to where the metalled lane became a grassy track leading off towards the fields. I said it must be a great place for stargazing, and she told me yes, it was.

We reached a chalet, a car on a gravel bed beside it, a small walled garden; the wall made of those cast concrete blocks in the shape of a flower. The shingle beach butted right up to the garden wall.

The door opened and a woman came out, followed by a man with a bucket. They greeted Marie and she told them about our meeting, just back there in the lane, after all those years.

'I can't believe it,' we said again, laughing, and looking into each other's eyes.

We reached a gate in the hedge, held fast by winter clematis and ivy. And like Mr Beaver disappearing into his dam, Marie vanished into the opening. I followed, curious.

There was a narrow garden passage leading to a front door. Suspended from branches, clipped onto fixings on the wall or standing on the gravel path, were dozens of blue pots with last summer's plant growth cut back to twig forms. Overhead, hawthorn trees spread their branches to form a private, half-wild, half-domesticated covered passageway.

We talked of meeting up again and hugged each other and then it was time to go.

I walked along the shingle beach towards the place I'd swum with Marie. I came to the rocks that Bill had carved and there, after so many years of weather and tides washing them over, his carvings, still legible:

Thank you Baycliff
for all the wonderful summers
I have enjoyed during my vocation on this planet
The year is 1972 AD…
Socialism is rife
In an unchristian society in which some people cannot afford to work and others have highly paid unnecessary jobs.

Someone had re-worked the word 'rife' so that it was now 'life'.

Elvis Presley King of rock music dies
Bing Crosby a great…
His signature is…
But the popular song of the year was
By the Swedish group Abba
The title Knowing Me Knowing You
I will write you the chorus

And he did…

Best wishes for you people
In 2077
Still civilised…
Us boys and girls…

Then stars awarded for all year-round swimmers, together with their names. Whilst the carvings are a rambling collection of thoughts, they clearly mattered, and were close to Bill's heart. Even now, if I hear the Abba song I have a small inward smile to myself.

Bill was in no sense an academic philosopher, but his words and messages play a role in maintaining the idea of the great English eccentric. Too often we don't think to record until we're past the moment of reckoning, and things that we don't regard as ephemeral prove to be just that.

I walked further, looking out at the bay. A pair of shelduck probed the sands. The light was changing again. Although it was still early afternoon, already there was a sense of things closing in, of the poignancy of winter afternoons when we want the light to stay, but we know full well that it won't.

Fifteen

Foulney – The Island that Wasn't There

'DON'T BE SILLY. There's no such place as Foulney Island, it's a product of your over-active imagination,' said my friend Ian years ago. And he'd been born and brought up in Ulverston, so he should have known better. He'd driven home from Barrow by the coast road one evening to see for himself. Then he'd called in to see me, admitted defeat. (I like a bit of 'told you so' every now and then.) But in a way, I thought years later, he'd had a point.

Foulney is always there, of course it is, but with its highest point standing at just three metres and its surface entirely flat and grassed over, it could easily be missed. And Ian isn't the only local I've met who didn't know the island existed. From the road it's just a raised grassy shingle bank, and at high tide there's not an awful lot to see. You might not even think to call it an island if you drove past on the way across to Roa Island and saw the bank sweeping away out into the bay. But go out there and see...

* *

November came in like a lamb. Still no frost to speak of but at last a break from the dull, damp final weeks of October. I hate the days without breaks in the cloud cover, I never get used to it; I feel hemmed in, tied down. But Callum was home from Edinburgh for the weekend, and checking the forecast for Barrow I'd seen that it was good. I checked too by looking out from the hall window and looking south and seeing the lighter sky.

We drove over, passing through showers as dark as thunder and on to clearer skies on the coast road, and parked up just off the main causeway to Roa Island. As we pulled on our boots and jackets a story came back.

I had friends from north Cumbria who lived up in the wilds above Brampton. They'd converted an old ambulance into a camper, and drove it down for the Town Hall Centenary celebrations in Barrow, an event that included the hoisting of a giant pair of Queen Victoria's undergarments from a flagpole above the Town Hall clock. That evening, as everything was winding up, Mike asked me to suggest a place to park for the night, somewhere they could wake up in sight of the sea, and I'd said Roa Island. Days later Mike phoned me.

'We found the car park OK. No one else there; not surprising really seeing it was already dark. We got settled in, got to sleep, but I needed a pee. Got up in the middle of the night and found myself standing in water at the bottom of the steps. The tide had come right up to the edge of the road!'

He just hadn't driven quite far enough.

'I got the van out of there alright, but it was bloody lucky I needed a pee, that's all I can say!'

Just as well. That might have been my credentials as an incipient tour guide shot to pieces.

Cal and I set off to walk to Foulney an hour before low tide. The causeway to the island was built by the Victorians as protection from the tides. It's a curving mound made of rock and shingle and concrete, and there are great gaps in it where the sea has torn chunks away. We picked our way along, looking down at sections of perforated black rocks like volcanic tuff.

At the car park it'd been warm and we left behind our hats and gloves. The further we walked out, though, the wind grew stronger and colder and we zipped ourselves tight inside our jackets. I stopped to look through the binoculars at something moving out on the sands. Into the lenses rolled a large white ball, a child's beach ball. We wondered if it'd blown all the way over the bay from Morecambe, what with the wind coming from due east. A solitary egret flapped out from the saltmarsh towards the distant sea.

Over time the island moves and shifts. You can see it by looking at the line of the causeway, and if it were thought of as the island's spine, it's as though this had been picked up and moved, transplanted into a different part of the body of the beast, but of course it's the island that moves and flows around the causeway. I liked this sense of an island as a dynamic, ever-shifting structure.

Once on the island itself we saw how it had been formed out of flat shingle, mountainous quantities of it, swept up towards a flattened crest where the tides had lifted the stones and re-deposited them in successive waves. Pity the poor wading birds nesting here, I thought; just one high tide and all evidence of their nests would be gone.

We walked up onto the grassy plateau passing close to the high-tide line of sculpted shingle, mussel shells, blue polypropylene rope (why always blue?), crab casts, deposits of blackened seaweed, plastic milk bottle tops, dried grass, twigs and dried plant remains, and far too many plastic bottles.

A procession of 12-foot-high metal poles led down the island, leading past a white tower that looked as if a rocket had crashed vertically into the earth. As we came closer it morphed into the base for a shipping-light orientated towards the outer limits of the bay. Further on, a concrete shed with most of the lower panels bashed out by the ravages of weather. Cal went inside and started to read the graffiti that was sprawled across the bare interior walls, and all that was visible of him were his legs and feet. I went inside too and hunkered down to look at two large blue fishing crates packed to the gunwales with large flat pebbles, each painted with a code and a number. OC 57, AT 48, RP 16, Lt 27, and then I got it; markers for the nest sites of oystercatchers, arctic terns, ringed plovers and little terns, left here by the Cumbria Wildlife Trust wardens after the end of the breeding season.

Once home again I'd looked up how successful, or as it turned out, unsuccessful, the season had been. My favourites, the little terns, had started well with 15 nest sites occupied, but a high spring tide had seen to them. If they were people, I'd rail at them for being so stupid: 'Just go and nest somewhere a bit further away from the sea, eh?'

The little terns decamped over to Walney's west coast, but no chicks were seen. There'd been 56 pairs of arctic terns settled on Foulney, but again no offspring. Predation by foxes is a factor of life and death for the ground-nesting birds here, and it was a canny opportunist who'd cleaned out all these eggs and chicks. I thought of the she-fox I'd seen at Aldingham after the spring gales and storms, and though I've always been a defender of the underdog (or fox), I know how I'd feel if I were to see it nest-raiding. At least the wildlife trust has new funding and a five-year plan to improve the island for the ground-nesters. I made a mental note to come back in late spring and see the island alive with birds.

A better adapted species, the eiders, fared much better. Morecambe Bay is the most southerly range of breeding eiders. Wardens had counted 170 chicks out on the water 'rafting', where adult birds gather the entire colony's chicks together and care for them collectively, both for safety and to teach them survival skills. As we walked on, in the far distance at the island's end, I saw eiders winging past in pairs or in small groups.

Cal and I began to wonder if we'd ever reach the end. Ahead the shingle had formed into two banks that extended out, wide from the central spit of land, like arms lifted at right angles to the body. To the right of us waves broke quietly at the low-tide edge of Walney Channel. A man in fishing gear walked rapidly along the edge of the sea. He was walking and stopping every so often to pick something up, but what, we were too far away to see. A Land Rover appeared moving over the sands exposed at low tide off Walney Island. It was being driven towards Piel Island, but once we'd taken our eyes off it, when we looked again it'd disappeared from view behind the island's only row of terraced cottages.

Gulls lifted and settled out at the far end of Foulney. As they rose upwards, the sun caught the undersides of their wings, gilding them like hot white metal. They glided, surfed air, drifted.

At last, we came to a slight rise and what we thought must be the end of the shingle bank, but it wasn't so. At this very low tide a swathe of rocky mussel beds extended out at least as far again as the length of Foulney. The ground was rocky and firm and we walked part way out to where the sea played, still distantly, coloured aquamarine, and where birds landed, foraging on the newly exposed mussel beds as the tide continued to retreat.

'It's like the moon,' Cal said.

And it was, an utterly grey, or colourless landscape and something, we both thought, we'd never seen before; land that had had the colour sucked out of it by the heave and flow of the tides. Two towers of red, scaffolded metal, presumably to mark the extent of the mussel beds to shipping, had been left high and dry, beached on the lunar landscape, like forgotten remnants from a film set.

There had been a simple shelter at the end of Foulney; I'd seen photos of it recently on the internet. A drystone arc of cobbles to keep the wind off. But the structure had disappeared, must have been sabotaged by the storms that hit the bay in the spring. Bizarrely though, on the barren shore great chunks of a demolished red-brick building had been scattered (if you can use the word 'scattered' about weighty chunks of a building). Whatever it had been, it'd had glazed white interior tiles as if from a school toilet.

This was the north-western extremity of Morecambe Bay. Given the state of the tides I doubted if the sea ever dropped back further than this, so the waters seen from here was the place the bay fused with the Irish Sea. In the distance a ferry ploughed steadily through the deepwater channel in towards Heysham.

Group after group of eider came winging over the channel from Walney Island, and with the wind behind them they moved effortlessly and at great speed. I wondered what that must feel like and I envied them, covering that distance, and being able to get right out there at the very edge of things. Following their drift we saw that at the very far end of the northerly 'arm' that pushed north into the bay, there was a row of what appeared to be ancient, blackened fishing nets strung out. How far away? Perhaps another mile or even two, the distance difficult to judge. The birds did it in seconds.

By then the cold was enervating. I needed to get out of the wind, to stop it bashing me around the head, and we turned and began to walk back towards the shingle. At the top of the beach, small cliffs had been scoured out by water and weather, no more than a couple of feet high. We stopped to take photographs of them. Here was the evidence, were it needed, that the whole island is made up of nothing more than a monumental stash of pebbles, all of them flat and rounded, and all stitched together with a scant mortar of soils and sediment and the penetrating, binding roots of grass.

Between Roa Island and Walney, windsurfers skimmed and bounced along the water. The wind was gaining strength continually and whitecaps were lifted out of the water. As we came back in sight of true land, another egret lifted from the marsh and disappeared out towards the middle of the bay.

⋆⋆

Collapsed and rusting at the side of the causeway road to Roa Island, there's a wrecked trawler. We walked around its bulk, taking pictures and wondering how it had come to be here. Someone had scraped back the rust and begun painting its hull bright yellow, though the job had never been completed and the newer paint had been bitten into again. Back home I found photos of the ship on the web, and the story of a family who'd taken it on and were making it into a home – of sorts. I guessed they'd abandoned her for a less arduous life by the time we were there. They'd called her *Vita Nova*, which had a kind of extra poignancy attached to it, the ship having had the life taken out of it twice. It must have been an adventure, without the high seas but with some of the romance. I couldn't imagine how that would be though, living your whole interior life at a tilt. I find normal interior life challenging enough.

I studied the Ordnance Survey map to understand the lie of the land around Foulney, and there on paper the extent of it was clear. Foulney itself becomes changed, no longer an island but a pathway out to the huge swathes of mussel banks exposed by low tides. There are even place names, the litany of each of them a story. The ridge where we saw fishing nets Slitch Ridge. Further north again, Long Barrow, Long Barrow Ridge, High Bottom, Blackamoor Ridge, Conger Stones. The scaffold lighthouse we'd walked out to, Foulney Hole, and I now see that it lies a mile further out from the end of the island. Ragman Ridge and Farhill Scar lie in towards the Walney Channel. Further out again Foulney Twist; sounds like a kind of local tobacco.

Then three more features further out into the bay caught my eye as it travelled over the map. The first a tiny scar, St Helena, named I could only think as a reference to the demise of the Emperor Napoleon. Further out again, another mussel bed named South America. But the next one unimaginable in a million years. A tiny scatter of rocks and mussel beds further out towards the extending finger of Mort Bank sandbank, named The Falklands. And I wondered if this had been noticed by any of the Argentinians who'd been here in the late '70s, working together with my father in the shipyard, before we went to war.

Sixteen

The Smallest Kingdom

AT NINE O'CLOCK in the morning the phone rang.

'It's John Murphy here. What's the weather doing where you are?'

'It's snowing – coming down pretty steady. What about where you are?'

'It's good. It was snowing earlier but it's clearing all the time. We're good to go.'

It was 16th January. I'd been trying to get across to Piel Island for four months without success. At last I was going.

* *

We drove seven or eight miles down the island road, and at Snab Point the right of way across the sands to Piel begins, a roughly semi-circular route that swings north towards the deepwater channel before cutting in a south-easterly direction towards the island.

John said, 'The pub won't be open today. But I phoned my pal Keith. He got a new coffee machine off his son for Christmas and he's very keen to try it out on us. We'll have a warm-up and then I'll take you round the island.'

At the beginning of the route, deep tyre ruts cut a swathe through the saltmarsh, gravel giving way to silt, saltmarsh giving way to the tidal bounds. The sign at the start of the route proclaimed: 'Danger, Soft Sand and Incoming Tides. No vehicular Access to the Foreshore'.

We were following the tide out. When we met that morning, John said we might be too early, that we might have to wait a while and see. But the first part of our journey was already clear of water, the sands sculpted by the tide into low elliptical banks and puddles and channels draining out from the land. In the distance a vehicle moved very slowly away from Piel, travelling through a sheet of seawater. It followed a route wide of the land, moving towards the main Walney Channel before curving back in towards the shore. To the east, way beyond the travelling car and further than the wide reaches of the bay, the land was under snow. It appeared to be held fast under the fresh fall all the way from the coast to the Pennine hills.

'You'll need to take care, especially over this first bit.' Under foot the first metres of the crossing were greasy and skiddy. We walked in the ruts left by the few vehicles allowed to travel the sands, then we were through and could walk freely again. I thought I'd dressed adequately for the job but in minutes of being out there I stopped to put on a windproof jacket. This was serious cold.

I had a walking pole as John suggested, and my notebook and pen, phone set to voice record, but the cold had already made my hands useless. I stuffed all but the camera into my pockets. Keep it simple, I thought. I felt a slight sense of unease as well as excitement. I suppose this was nothing more than a healthy self-regard, me in survivalist mode, setting out into what was for me, the unknown. But John has been guiding walks to Piel for 20 years, and his enthusiasm was reassuring.

'We're away, we're OK,' he said. 'We've got a brilliant day for it. This is going to be good.'

★★

It's not every day I get to email a king, but the landlord of the island's pub has been known as the King of Piel since one of the more obscure and bizarre episodes of English history. The story centres on the accession to the throne after the sudden death in 1483 of Edward IV, the first Yorkist King of England, whose two sons, Edward and Richard, were subsequently imprisoned in the Tower of London by the Duke of Gloucester. A young boy born of humble origin, and who became known as Lambert Simnel, was discovered by the Oxford-educated priest, Roger Simon. Simon claimed the boy bore a striking resemblance to the two princes in the Tower. Simon educated the boy in courtly manners, and initially planned to present him as Richard, Duke of York, the younger of Edward's sons. When he heard rumours that Richard had died, he then presented his protégé as Edward V, heir to the throne, who had miraculously managed to escape.

Simon took Simnel to Ireland, garnering support from the Irish government, and at the age of 10, the boy was crowned King Edward V in Dublin Cathedral. An army of Flemish, Irish and a number of English troops was assembled. They sailed for England and landed on Piel Island on 5th June 1487, where Simnel immediately laid his claim to the English throne. The rebellion was quashed as Lambert's army moved south and was overthrown at the Battle of Stoke Field. Simon avoided execution because of his priestly status but was imprisoned for life. As for Simnel, he was pardoned by Henry VII for being nothing more than a puppet, and given the position of spit-turner in the royal kitchens.

Later he married and become a falconer. One can only goggle at what must have gone through the boy's head, educated out of his simple early life, given two regal identities within a short amount of time, and paraded as the rightful king at the head of a rag-tag army. I wondered though, if he ever thought of Piel, and what he'd made of it, or if he might have imagined his falcons flying above the island as he watched, a hand shading his eyes from the too-bright sun.

This inglorious attempt at kingmaking led centuries later to the gently mocking and bizarre custom of each incoming landlord of Piel's Ship Inn being crowned as the King of Piel. The new monarch sits in a battered, ancient wooden throne wearing a helmet and holding a sword whilst he is anointed by the pouring of bucketfuls of beer over his head. Graffiti carved into the chair's pitted surface date the ceremony back to at least 1856, and the care of the helmet and throne are an integral part of the tenancy agreement.

In the early 20th century there was at one point an entire cabinet that included a Prime Minister and the Lord Mayor of Piel, together with a royal family. Knights (and there are women knights too) are created by the same process of induction and are then said to be free men of 'the Noble Ancient Castle of Piel'. And just so we're all clear, the king and each knight was expected to be 'a free drinker and smoker and lover of the female sex'.[24]

★★

I'd visited the island previously, but not for years. The last time it had been a family day out with pals. Steve and I took the ferry across from Roa Island with the kids and dogs and the other adults took off in a Canadian canoe.

'I hope they know what they're doing,' the ferryman said, mid-stream, 'I'd not like to attempt that in an ebb tide. We'll never see them again.'

As soon as we'd landed on Piel the friends' dog found a fish head and set about eating it, together with the fishhook that was still in place, and that then became stitched into the dog's lip.

Thank God for mobile phones – we called our vet friend and he talked us through what to do. It made for a memorable visit. We'd had a drink at the pub and taken the kids to the castle and I remember looking over the channel towards Walney and seeing a hawk moving towards us. At first I'd thought it a kestrel. As it came closer, a small flock of starlings whizzed over the castle bailey and pinged apart, scattering in four directions. As the hawk swooped and dived after its prey, it passed close by us and before it disappeared I saw that it was a sparrowhawk.

I've had two more recent close encounters with sparrowhawks. Involved in the mundane task of washing up, I'd looked out of the kitchen window and seen one perched in the top of our clematis-covered trellis. The hawk was clearly watching me. I called Steve and together we looked back; the bird was a mere five or six feet away. Over the course of the next minutes, it didn't once remove that piercing yellow gaze from us. It was as if it had come there with the purpose of watching our every move, and I think we both felt a slight frisson of intimidation.

Another morning I'd been setting out with the dog and seen a sparrowhawk winging in from the direction of the fell. I knew that my neighbours from the allotments up there were incensed by the things; the hawks patrolled the fell regularly, looking out for the relatively easy pickings of large numbers of racing pigeons. About 50 yards down the lane, on the hard standing in front of a neighbour's garage, the sparrowhawk stood, one talon pinning a pigeon's wing to the ground. The pigeon was on its back and, with its other talon, the hunter pounded up and down on its victim's chest. The dog and I stopped still. The sparrowhawk locked eyes with mine; again that intimidating stare. It continued pumping the pigeon's chest, driving its claws in, trying to compress and stop the bird's heart. The hawk's stare had about it an air of unquestionable superiority. 'Interfere if you dare,' it seemed to say.

I didn't, and I don't think the dog was keen to linger either. We left the scene of the crime. Coming back from the walk there was no sign of either bird, just a collection of pigeon feathers littering the ground.

★ ★

Planning my return to the island at the beginning of autumn I emailed the landlord of the Ship Inn to ask about ferry times. When the answer came, it was not what I'd hoped – or expected.

'The ferry's suspended until the spring. Sorry.'

I wrote again. I told them I needed to get to the island to complete writing a book.

'The Roa Island jetty is being rebuilt. Consequently the ferry service is not able to operate.'

'Is there any other way to get to the island?'

'The only way is to walk, but the sands are treacherous at the moment. Boat is not an option as jetty demolished so no public access. Sorry.'

I began to search for alternatives. I found a name and a number for the ferry operator himself. I phoned. We talked about the possibility, the logistics, but despite the non-existent jetty it seemed we had a plan.

'I can pick you up from Roa Island, but you'll need good, high wellies and be prepared to get a bit wet.'

Then the evening before, a message came through: 'Ferry won't go. Think there's water in the block. Sorry.'

Piel was beginning to feel as remote as St Kilda.

Just after Christmas an email came through from Sheila at the pub: 'Hi Karen, there's a walk on New Year's Day across the sands to Piel accompanied by the local guide, John Murphy.' I found it in my spam folder a week after the event. But I had a name and in minutes I was talking to John on the phone. At last, I had my man.

**

Mention the term sand-pilot and Cedric Robinson's name will invariably crop up as he's the celebrity guide, favoured by royal appointment. Cedric leads walks and events that take in bay crossings from Kents Bank across to Silverdale and Arnside, with slight variations in destination according to the tides and the condition of the sands and the channel.

But the historic longer nine-mile crossing from Hest Bank to Kents Bank, crossing the channels of the Kent and the Keer, is currently not being walked. Stephen Clarke and Alan Sledmore led this arterial route using quad-bikes, and if smaller children became tired, they could have a turn on a seat on the back. But the Council weren't happy, and the walks stopped a while back. I asked Lancaster City Council if there were plans for guided walks in the future, but they didn't seem to know and directed me to their tourist office, and they could only tell me about Cedric's walks.

It seems unthinkable that the route will remain abandoned. With Cedric now past his 80th year and keeping to the part of the bay he knows the best (quite rightly, too), the long crossing is in danger of becoming the stuff of legend.

The crossings of Morecambe Bay are neither pathways nor sea-routes. They are hard won negotiations with sea-swells, 10-metre tides and storms. It is not possible to create foolproof waymarkings because of the shifting nature of the sands and the river channels. As Turner and Cox painted laurel brobs set in the sands, so the tried and tested method of temporary waymarking continues. Whilst technological mapping can mark routes to a hair's-breadth accuracy, GPS would be useless; it wouldn't recognise quicksand, the approach of the tide or the depth of the channels.

And there's a conundrum. How could you begin to train sand-pilots? How many people have the necessary knowledge, or would want to take on the responsibility in this day and age? Ask locally, and it seems there's no one waiting in the wings. Unless there's already a background layer of experience in the bank, it simply isn't accruable.

The less well-known crossing of the Leven sands from Flookburgh to Ulverston, led by Raymond Porter, historically took in Chapel Island and was frequently used as a refuge by travellers caught out by the tide. In the summer there are guided walks across to it, but even that's not always possible. When I talked to Raymond in the autumn he'd said the sands had been unseasonably difficult because of heavy rain. Despite that I came across a geo-caching website telling me that it was a great trip, and that you could walk across at low tide. Oh, if only it was that simple.

If George Stephenson had had his way, Chapel Island would have become a railway station. He was investigating alternative train routes north to the hilly route over Shap, where the West Coast mainline to Glasgow was later constructed. Stephenson's plan was to take the railway all the way across the sands from Morecambe to Humphrey Head on the Cartmel peninsula and then on again crossing the Leven Estuary into Furness. Imagine that.

The idea of a bridge across the bay has been talked about for years. My father was a huge enthusiast. 'It'll bring investment, boost the economy of the area. Imagine – it'd only take 20 minutes to drive to Morecambe,' he was fond of saying. The same reasons are wheeled out every few years whenever anyone has a go and gets the local press jumping with a project that will never happen. At Lancaster University Dr George Agidas has researched the possibility of a road-bridge that incorporates tidal-powered turbines. His studies pinpoint the current locations in the bay where turbines would be placed to give maximum output. Inevitably, any scheme like this would invoke heated debates between those in favour and those with an eye on both the visual and environmental impacts. And what if the bay makes its own decisions? Who's to say where the area of greatest tidal push and pull will be found in another 30 years?

* *

It wasn't every day that one could boast of having walked to an island. Once we had set off from Snab Point on the shore of Walney Island and manoeuvred across the slippery beginning of the walk, we were out on the open sands. I turned round to orientate myself. Further up towards Barrow the channel widened. Walney was reduced to an elongated, undulating strip of land. Snow showers bloomed over Black Combe, moving north over the Furness hills and travelling towards the mountains. We watched a jostle of geese coalesce into an organised form following the strip of land south and out into the Irish Sea. Then John pointed out a dark shape in the sands.

'Someone obviously thought they knew what they were doing.' Way out towards the Walney Channel, a car, buried to above window height.

'There's another wreck further on – we'll go past it. We'll call in at Sheep Island first,' and minutes later we walked up onto the beach of a small island a short distance offshore. We crunched over the shingle. It seemed the island's only features were a solid fence constructed at the line of vegetation and beyond it a dense tangle of plants.

As we stepped back onto the sands John listed the names of the nine islands of Barrow. Most of them I'd known, some not.

'Walney; Roa; Barrow Island; Crab Island in the middle of the channel, also called Dova Haw; then Sheep Island; Foulney, named after 'fowl' for its nesting birds; Ramsey Island, which became incorporated into the dock wall; then Piel; and last, Headin Haw.' The last just a speck of a place offshore from the docks, once used as a dynamite store.

'See the edge of Black Combe? In the last Ice Age the glaciers were heading south and bulldozed the face off the mountain – see where it looks dish-shaped? The deposits were dumped here – making all these islands.

'A long time ago shipping used this side of the channel, coming into port this side of Piel.' John's arm swept up along the sea of sands that stretched between Piel Island and the inland coast of Walney. 'Then the sands and the channel shifted. In the 1800s there was an isolation hospital on Sheep Island, built out of wood.'

'Must have been draughty,' I said.

'It didn't last long. It was supposed to be for sailors coming into Barrow who might have contracted contagious diseases. I don't think it was used much though. It closed in the 1920s.'

The four-wheel drive was coming closer.

'That's the King,' John said. The vehicle pulled up beside us and the two men exchanged greetings.

'I'm off for gas bottles,' the King said.

By now the morning light had grown in intensity. Overhead the complex white masses were drifting apart and sunlight fell onto the wet sands. More snow – or hail – fell on Barrow, which, being so close to the sea, doesn't happen too often. It fell in a series of chaotic bands as if the wind behind it had no idea which way to blow. Away out on the Irish Sea, great dark banks of cloud broke apart, and the light that fell from the sky super-lit the sands under our feet. Striations of linear markings had been formed by the retreating tide, and these, together with the intensity of the light reflected on the sands and the speed of the wind, made it seem as if we were turning under the sky, and not the clouds moving over our heads. The snow showers pushed out into the wide open space of the bay. It was a sky that would have had David Cox rushing for his boots and paints.

'Look there,' John said, 'a rainbow being born.' And underneath the edge of falling snow, high up into the far reaches of the bay, the unmistakeable shape and colour of a rainbow, bright against the black sky beyond.

We came alongside the remains of a van. Had it not been for the steering wheel, it might have remained undecipherable, and more like a whale carcass fashioned out of rusting metal. Its whole form was garlanded with weed of the deep sea-greens, reds and pale ochres of the sub-sea world. Each rusted pinion and shaft was carefully wound and transformed. The tips of two rear tyre arches broke from the ground, the prop shaft bent in the centre, broken in two.

'That one was a Sherpa van. There was another one not too long ago. Keith will tell you about that. That bloke was very lucky to get off at all. There've been a number of drownings over the past few years. Six, I think.'

One of the windfarm boats slunk seawards down the channel. It was a strange sensation, of walking on the sea bed eye-level with that wall of steel as the ship slipped towards open water. In dark silhouette Piel appeared more like an abandoned stage-set than an island. To its north a gaunt terrace of six houses, then the pub, whitewashed like a traditional Lakeland farm cottage, red chimney pots a-crest its slated roof, and the jumbled ruins of the castle facing out to sea. We were coming closer, then the sound of a car again and the King reappeared alongside, one arm resting on the window frame.

'No gas,' he said with a smile. 'No matter, we'll manage.'

And as he drove on, John said, 'He's a great chef, a really good chef,' and the King's car crunched up onto the island's foreshore. John told us about the walk he'd led over to the island on New Year's Day, with mulled wine and a good fire in the pub. It sounded a perfect way to begin the year. After wading through an ankle-deep, narrow channel of standing water, we landed on Piel. Thank God for Gore-Tex boots.

* *

'This is Keith, this is Karen, and that's Charlie the dog,' John said, introducing me to the island's only full-time residents other than the King and Queen.

'I'm Karen too,' I said. We left our boots outside and stepped into a small sitting room. The warm blast from a steeply banked-up coal fire soon hit home and we shed our windproofs and jackets. The walls were painted red and the ambience of the place was one of warmth and homeliness.

'We were expecting you, glad you made it. Now let's get down to business – coffee?' Keith said. 'I can highly recommend the caramel latte macchiato.'

I'd just walked over the sands to a remote island with only four inhabitants. I'd not been as cold since I'd been lost in a white-out on the top of Ben More Coigach in Sutherland over 20 years ago, and now, with my cheeks beginning to turn red from the heat, I was being offered a coffee menu to beat the high street chains hands down. A few minutes later we were handed coffee glasses elegantly striped with layers of caramel and cream and coffee.

Despite the strong wind and the cold, the house was warm and solid. Paintings with maritime themes decorated the walls: clippers surging through the sea, small boats at anchor. Karen told us about their return to the north, how they couldn't wait to get back after 26 years living and working in the deep south. The house had belonged to Keith's mother, and when the chance to take it on came, they'd jumped. They moved in during the winter of 2010, and the day after their move it snowed – unusual for Barrow and unheard of for Piel. And the snow froze, and the sea froze too and the ice remained for a month. I recalled that winter too, going to Sandside on Christmas Eve to witness the frozen sea, walking the saltmarsh towards Arnside and the afternoon's lowering sun casting our shadows like slim angels over the incoming tide beyond the edge of land. And where the water had pushed into the channels that furrow the marsh, what remained were intricately detailed, frozen, upright fans of ice, for all the world like starched Elizabethan lace collars. I recall the sound of it too, the tide pushing in a great slow, slushing, chinking curve from the far side of the estuary, directed through ever-narrowing strips of semi-open water. And icebergs, for goodness' sake; small, yes, but nevertheless, icebergs in Morecambe Bay.

* *

Keith has island blood in his veins. His grandfather was King for many years, and Keith grew up in the pub. As a lad he understood that everything had to be carried across the narrow stretch of water between Roa Island and Piel – barrels of beer, crates, fuel, food; everything. In late autumn two boatloads of sheep were crammed onto the open ferry, 70 at a time, to overwinter on the island's grass. Keith began operating the ferry himself at the age of nine, and by 16 was granted his boatman's license. There had been no phone or radio, no way of making emergency contact with the mainland other than improvising, and on the night his grandfather suffered a major stroke, Keith's mother ran to the slipway with a piece of fabric soaked in paraffin, set it alight and waved and waved and waved until eventually, after too long a time, a light of recognition came from the other side.

'We used to run the pub for two weeks every autumn so that the last landlord, Rod, could have a break. We were lying in bed one night. We knew there was absolutely no one on the island apart from us. Then three loud knocks came on the bedroom door. We both heard it, looked at each other and decided we were not about to go and see who was there!' And Karen had been aware of the presence of male figures at times, wearing old-fashioned uniforms like the pilots and coastguards, and once she'd asked Keith why he hadn't answered her question – but he'd not been in the room at the time.

They seemed so composed and happy, these island-dwellers, comfortable with their world. But island life has its challenges. In mileage terms it's not that remote, but given the tidal nature of accessibility, you might as well be living in the Hebrides. Karen has had health problems, but is much improved and actually looks the picture of health. But then Keith fell, breaking his ankle and shin and tearing ligaments, and Karen doesn't drive.

'Luckily we had an old school friend staying at the time. He's the Dean over at Lancaster. We were about to set off across the sands to take him back. There were big tides on and a Force 9 to 10 blowing in. The others were already in the car. It was so cold I set off running over the wet grass and a big gust slammed into me, took me off balance and down I went. I heard the crack – I knew it wasn't good. I passed out. I just remember coming round in a large puddle, the water up to one lens of my glasses, and thinking, "This water's very cold." I didn't know it then but I'd broken my ankle in four pieces, smashed ligaments and broken the shin.

'It's no joke when the ambulance service doesn't have the foggiest where you are. The operator had never heard of Piel Island or Snab Point. Our friend Steve had to drive the pick-up. They got me in the front and I remember thinking, he's heading too close to the channel, and having to direct him. I managed to keep it together till we met the ambulance, then once I was onboard and Karen and Steve were away in the car, I couldn't get the pain relief quick enough.'

'How on earth did you manage living here?' I asked. Their two sons had come north to get them sorted; bagging up several tons of coal so that Karen could carry small amounts into the house at a time. There's no mains power on the island, and shortly after the accident, their generator packed in. With the other five houses in the terrace run as holiday cottages, and a total full-time population of four people, there's not always someone around to lend a hand.

'It's not so bad these days though: Tesco and Asda deliver to Snab Point. We just have to drive to the other side to pick it up.'

'They're not offering to deliver to the door then?' I asked, half joking.

'Not yet. A healthy regard for the sands is a good thing.' Then, 'A young lad got his car stuck, back-end of last year. He came knocking on the door late one night. It was October, so it was pitch dark. I took him back out in my car, and found a 16-year-old lass in the car, wearing a dressing gown. She was in bits by the time I got out there. I tried to pull his car out with mine, but it didn't budge – my rear wheels were lifting up into the air. Then the fuel pipe snapped, so that was my car written off. He was lucky though, that lad – and his girlfriend – with the tide on its way in. I'll hand it to him though, he came back next day and dug his whole car out. It'd sunk halfway down the spare wheel on the back door. Then he towed mine off for me.'

John and I had to keep an eye on the time, and although it would have been good to stay longer by the fire and listening to island stories, we pulled on our boots and warm gear and said our farewells, and Keith said, 'When you go round, look at the back of the terrace and you'll see a tiny window high up on the back wall of each house. There were stairs up to them – we found the remains of them when we renovated the house – so the pilots could keep look out for the masts of sailing ships appearing over the top of Walney, then they'd race out in rowing boats to guide them in.' I liked the image, the romance.

'What do you make of the turbines though?' I asked.

'I don't mind them, actually. They make me think of *Don Quixote*. At night, when the service boats go out and they're all lit up, it's quite a sight.'

* *

Someone had made a sign for the island campsite. It was tilted at a jaunty angle atop a grassy mound: 'Benefits Street'. Good old Barrow humour. We walked the shingly beach towards the castle ruins, John stopping to show me Sailors Soap seaweed and sea-rolled and rounded nuggets of red-brown iron, remnants from an ore-laden ship that had been wrecked offshore.

The castle covers an impressive area. It's had the usual kind of restoration, the newly carved sandstone blocks apparent against the wind-worn ancient architecture. Built by an abbot of Furness Abbey, Piel Castle has long been considered the storage facility for an illicit monastery trade in brandy and wool. With an outer moat and fortifications, the monks were clearly serious about keeping their goods out of sight and safe, and keeping potential raiders out. Here on Piel, there was of course ready access to trade routes, but also it was strategically positioned to guard the deepwater harbour of Barrow-in-Furness against pirates and raiders from the north. I remember stories I'd heard years ago, of a tunnel connecting the island with Furness Abbey back in the hills beyond Barrow, the once fabulously wealthy stronghold of the abbey set deep in a natural amphitheatre in the Vale of the Deadly Nightshade. And I thought of poor old Lambert Simnel holed up here, holding court, though God knows how he did; waiting to be told what to say and who to say it to, no doubt.

I watched a tawny-winged kestrel hanging in the air beyond the castle keep. It was spotlit by bright sunlight. Inside the shelter of the castle keep, we were protected from the buffeting wind, and could almost feel the slightest sense of warmth from the sun. The kestrel, superimposed against the whitest of clouds, eyed the earth, then fell from sight.

We circumnavigated the island. As the view eastwards over the bay opened out I saw again the snow-laden hills of Pendle in Lancashire. Another line of birds came in from the open sea, geese again, using the southern tip of Walney as the place to begin forming into an organised V, then flowing onto the land close to the shelter of the island's inland pools. And just there, the sandy bay where grey seals haul out, though they weren't in evidence today. A path led us under the castle walls and past a low cliff, underneath which lay a substantial portion of a tower, collapsed and improbable, built of a confluence of beach cobbles, bricks and mortar. The wind high-pitched itself from out of the ruins like the sound of 30 boys whistling, all in a different key. A complication of briar and hedge bordered the path and a blackbird landed ahead of us, tail up, cocky, regarding us with that sun-ringed eye.

'Look out there,' John said, 'Seldom Seen – you can see Seldom Seen.' And he pointed out into the gunmetal water to a dark strip emerging from the southernmost reaches of the bay, materialising and vanishing again and again as the sea washed it over and over.

'Mussel beds,' John said.

Oystercatchers came in off the bay, settling on the grass and in the chicken coops behind the pub. They looked at home, as if they knew exactly where they were. And as we looked out over the lands of Furness, the 'Far Ness' or strip of land that the Vikings named, the sky darkened to the colour of Cumbrian slate, and born on the body of the sea more white squalls moved, pushed on by air from the north, drifting like wraiths as they passed over the narrow band of Foulney Island, out towards St Helena,

South America and the Falklands. And out of nowhere another kind of cloud appeared, a gathering of knot, reeling and flowing over Seldom Seen in their micro-choreographed unanimity. Like a fish shoal they turned, changing colour immediately and in an instant, though we searched for them long after, they had gone.

Seventeen

Small White Ghosts of the Sea

Inside the hide there was shelter from the boisterous wind coming in off the Irish Sea. In the middle distance a small, sky-blue boat leaned into the saltmarsh where it had been left at the high-water mark. It was low tide. I looked over the inlet towards the final crooked finger end of Walney Island, a shingle spit formed of long-shore drift from the sea-rolled west coast. This was home to the largest gull colony on Walney. In the sky above it an aerial flotsam and jetsam of birds moved and flickered, glinting white, grey and black, melding into and out of the spun silver cloud that persisted over the bay. It was a whole society of birds, a sky-city of herring and lesser black-backed gulls that glided, drifting and weaving and passing through each other's air-drawings, hanging on thermals. The clamour and noise of them carried across the inlet towards me. Near the ground a heat-haze had formed and the young gulls that strutted between nest sites became fractal versions of themselves. As they wandered, their heads swivelled from side to side, looking around as if weighing it all up – this life, and this place.

It'd been too many years since I'd last been to the island. In the mid '80s I'd come with a group of schoolchildren. It must have been a week or two further into the summer as the chicks had hatched and fledged. Nest-sites had only the remains of marbled gull eggs and the downy feather nest linings were left for the wind to disperse. We walked down the island road, passing the lighthouse where Peggy Braithwaite then reigned, Britain's only principal female lighthouse keeper. We'd been advised to wear headgear in case the gulls dive-bombed us, and that day my colleague Martin was wearing one of Peggy's knitted bobble hats; as well as her lighthouse duties, Peggy knitted to raise money for the lifeboats. Years later, when I'd found out more about her, I really wished I'd met her.

* *

We had journeyed all the way down to the southernmost point of the island, to the place where water from the Walney Channel and Morecambe Bay assimilated with inward-moving currents from the Irish Sea, and where a flume of white water roiled seawards. In the air above it a group of little terns – or 'sea swallows', as they're sometimes called – danced the bright afternoon away. It was the first time I'd seen them, and first time for the bird-mad children too. We listened to their small voices, a rolling, squeaky call that mimicked their ever-restless nature as they flittered through the air, then came down to touch the surface of the water for the merest flick of time before rising again in a continuous wave of activity.

For once the children had been quiet. They were used to birdwatching with us, but there was something other-worldly about the terns. Swallow-like, they flew with the characteristic collapse and fill of the wings, and although larger than swallows they seemed less substantial somehow; I remember that the response 'ghost-like' formed in my mind. Then only last summer I'd talked to a friend who'd just seen little terns for the first time. She'd also described them as swallow-like, insubstantial and other-worldly; that they were like little ghosts.

More recently I'd read that the little terns hadn't fared well in the intervening years. I phoned ahead of my visit and asked the warden about them.

'They nest precariously close to the tideline, so just one big tide and their nests would be washed away. Then there are the predators; last year we had six kestrels hanging around, all hungry. There are foxes and gulls as well, and this year we've seen crows taking the young.'

'So it's not looking too good?' I'd asked.

'Who knows? They were absent from the island for a decade until last year. They do better across the channel on Foulney Island. There are fewer predators to bother them over there. We try not to create any more disturbance than is necessary, but sometimes we manage to put fences around the nest-sites to stop the foxes getting in. But that's not going to stop the sea washing them out.'

* *

From the hide I looked across to Roa Island where the new lifeboat station seemed anachronistic at low tide. It had been built on tall stilts and had the look of an oversized bathing machine, or else a house future-proofed against rising sea levels. Piel Island appeared adrift in the middle of a sea of sand, and the eponymous sandstone castle, worn smooth and rounded from centuries of weather, flared red-ochre in the morning sunlight. There'd been a ship too, sitting at anchor in the channel, glimpsed behind and through the castle ruins.

Then the ship began to move, travelling into my range of vision. It slid around the shingle spit, appearing improbably close to the land. Through the binoculars I saw men on the bridge, and others on deck, leaning on folded arms on the railings and looking down into the water as if they too couldn't quite believe how close they were to land. Skeletal red channel markers guided the ship out into the Irish Sea and it began to plough across the water towards the quadrants of a vast windfarm, the Walney Array, then under construction. I find the use of the word 'Array' interesting. It suggests an aesthetic plan, but to me the windfarm was anything but. I like my nature untrammelled and my views unrestricted. The turbines jarred, as if the horizon had been scratched into.

In the midst of the turbines a crane on a huge construction platform dangled a turbine tower over the water as though it was a toy. Then a small boat came zipping away from the turbines and made its way around the tip of the island into the channel and headed inland towards the harbour at Barrow.

I left the hide. The sky had begun to clear as the tide turned and began to fill. I'd noticed this conversation between the sea and the sky many times on my journey around the bay. Clouds were being created out of the uniform grey. They formed then dissolved and disappeared in minutes. Above them cirrus concluded in replicating curves where the wind spun them away into the blue. Curlew and oystercatchers flew overhead, passing from one side of the island to the other in nothing more than a few beats of their wings.

I moved on, walking in the lee of the wind. It was warmer and I was sheltered by high, grassy dunes to my left. On the right of the path a series of lagoons were populated by eider and mallard and moorhen. I tuned in to the sounds they made, the coo and twitter of the eider, snuggled on small grassy islands dozing the day away. Meadow pipits whistled and mallard and moorhen paddled themselves around. A large bee came towards me on the path, igniting the air with the fizz of a sparkler.

Passing the lighthouse and the cottages, I walked down to the west through the dunes. In the language of my childhood, the sea was far away, across a wide band of sand riddled with shallow sky-pools. Cormorants out at the water's edge were like dark sentinels looking out at the unnatural towers of white over a sea that barely moved.

The sky was full of gulls again. In our phone call the warden had told me that some of the gull-chicks had already hatched, so I knew I'd have to keep moving. I walked parallel with the coast on a track through the middle of the gull colony. The scrapes were everywhere, a few just a collection of downy feathers and the cracked-open remains of a solitary brown-black veined egg. Overhead the birds reacted to my presence, dropping out of the air towards me, clamouring and calling. In the few scrapes where chicks had hatched, the parent birds became increasingly frantic. I held my walking pole up into the air so that when birds took aim for me they targeted that instead. Where gulls were sitting on eggs, their heads popped out of the nests, swivelling so as to see what and where the trouble was. If I passed too close, they rose up into the air, keeping me in their gaze all the while.

The restlessness among the colony was contagious: pulse after pulse of anxiety flickered through them like a frantic Mexican wave. On the roof of an ancient concrete shelter, a remnant from the war, piles of wind-blown sand had become stitched together by marram grass. Two herring gulls stood upright, emerging from the grass as I walked towards it. Two pairs of yellow eyes gimballed, following me as I moved past. Camouflaged in the grass beside them, a single large chick that moved slowly and purposefully, picking one foot up before the next and moving into the shelter of the turf until it disappeared completely from view.

The stress my presence created became too much for one gull. It swung hard in to scare me off. I ducked and laughed as it banked and turned and came in again for another shot; I was being duly escorted off the premises.

Gradually the nest-sites became fewer until I'd left the colony behind and the gulls quietened again. A stocky pony grazed, tearing at the grass. I walked along in the gentler company of skylarks and a softening wind. The day was opening itself up and the light had become bright and clear. The sea sparkled as waves formed and broke again. Swallows wafted to and fro across the dunes, and in a few minutes more I saw the Sea Hide.

Even before I'd gone inside, there came, borne on the sea wind, the faint and unmistakeable fret and twitter of little terns. Once inside I opened up the windows and the breeze blew straight in from the sea, amplifying the terns' voices slightly. It took a few seconds more to find them, first by eye and then through the binoculars. They were there, a small number of them, down towards the sea-edge of the shingle beach, so few though that they could easily be overlooked. As I watched them they landed and lifted continuously in that particular, perpetual rolling motion of theirs. Watching the little terns, it seemed that their restless behaviour mirrored their vulnerability; this year just 15 pairs had returned to the island. I watched as they flitted between sea-pools in the sand and their nests on the shingle, intent apparently on feeding chicks, though they remained unseen, camouflaged amongst the pebbles and the beach colours.

There's an element of the terns' vulnerability that adds to the thrill of seeing them, but my reaction that day took me by surprise and I wasn't anticipating it. There was a definite shift of adrenalin and the instinct, though I'd learned to suppress it, to shout, 'There!' This emotional connection with certain wild creatures sustains and lifts us up, and it can be disarming. But there was something else too, about the way that time can suddenly collapse like that. It elided, sliding in from my life in Ulverston all the way to that moment; it left me questioning, and wondering. And who'd have thought it? All this from the sight of a few insubstantial birds flitting and fretting on the shoreline, those small white ghosts of the sea.

★★

Lighthouses. Just say the word, and it conjures up a whole world: charm, the stuff of children's stories or symbols of strength and indomitable spirit, of engineering wonders. And for me, of a lighthouse without any light, a mile away from the sea. But the story of one of the keepers here at Walney Island, the only female ever to have held the post of principal lighthouse keeper, remains unparalleled in British lighthouse history.

When I began to research her story I realised just what I'd missed by not meeting Peggy; she died in 1996 aged 76, and just two years into her retirement. In his obituary piece in *The Guardian*, the journalist Martin Wainwright described a redoubtable woman who 'ruled a unique kingdom on the shores of the Irish Sea' whilst 'dangling in a boson's chair'.[25]

Peggy was born on tiny, neighbouring Piel Island. Her father ran the small ferry between Piel and the mainland. When he was appointed assistant principal keeper for Walney Island, he moved his family and all their possessions across the channel in a small open boat. Peggy, her sister Ella, her grandmother and the family's piano were all crammed in on the final trip. In the absence of sons, the keeper trained both his daughters as lighthouse assistants. When her father retired, Peggy was promoted to principal keeper.

On an October morning I returned again to the island. I drove from Kendal in filthy weather. Great sheets of blue-grey clouds persisted over the mountains and extensive bands of rain obliterated vast tracts of the higher ground. But down at the island the sky began to clear, and when the sun broke through at last, light bounced off the sea as if reflected from metal. I had a rendezvous with the harbour master from Glasson Dock who came over twice a year to check the lighthouse and make repairs.[26] I'd been put in touch with Brian and emailed him about my interest in Peggy and the island. He offered me an opportunity to see the world from Peggy's point of view.

There was a peculiar foreshortening of distances that day. Waiting in my car, I looked due north over the channel. Strong sunlight highlighted areas of the landscape so that they appeared closer; Hoad monument glimmered, standing out from the background of hills as if a mere couple of miles away and not the 12 or so it really was.

When Brian arrived (yes, the third Brian in this book, but I do know men with other Christian names), we continued the journey down the island in his car, ploughing into and out of rain-filled craters, the car rocking from side to side as though a ship ploughing through waves. At the lighthouse a group of children were running in circles around puddles, oblivious to the cold. They were on a school visit and couldn't wait to climb to the top. The first small group was summoned and once inside and out of the wind, they paced up the steps, counting as they climbed: '… forty-seven, forty-eight, forty-nine, FIFTY,' the way they do.

We emerged into the rotunda and the light itself. Unlike the complex and bulky arrangement of mirrors I had expected, it was smaller, more streamlined, a set of four dished reflectors set inside a glass roundel. The Walney light was maintained by hand up until 2003, and was one of the final few to become automated.

The children sat down on a circumference of stone and began asking questions and discussing the light. Brian indicated the narrow door and suggested I might get better photographs out there. As soon as I moved through the door and onto the external gallery, the wind caught me unawares. It was blowing hard, a gusty wind that had me reaching into a pocket for my hat. Straightaway I saw the effects of centuries of long-shore drift, the washing out and re-depositing of stones by the sea and the gradual accretion of land. Though it was built at the southern end of the island, the tower and its once keeper's cottage now fitted snugly in the centre of a system of dunes and saltmarsh.

It was an oystercatcher's take on the island, on the world. Water channels cut through the saltmarsh in curls and winding shapes like a piece of Indian fabric. Offshore the windfarm was there, impossible to ignore. The shadow of the lighthouse pointed north-west towards the docks and town and beyond to Black Combe's bulk across the Duddon estuary. On the southernmost spit, the place where the herring gulls and lesser black-backed gulls of summer had reigned and filled the air, a small number of them moved like tiny shreds in the wind. The grey seal colony and their young were hauled out on the beach by the side of the Walney Channel. The other children and their teachers became smaller as they walked towards the hide on the island's western coast.

I looked down at the ground, at the cobble-walled gardens and chicken sheds. Being that high up and with the crazy, battering wind I felt the push and pull of giddiness. Looking over the edge of the guardrail, I thought of Peggy setting off in her boson's chair to repaint the tower – what a woman. I imagined her singing to herself as she cast the dangling seat out from the top rail, organised with paint, brushes and rags, surveying her island province and stopping once in a while to take in the view. Oystercatchers might have winged past so that she'd felt the breeze their wings made. She might have watched them bank and curve around the southernmost spit of the island before following the channel north again. I imagine she wouldn't have wanted to be anywhere else on earth.

Over the sea the light was growing brighter. It was hard to look without shading my eyes, but moving around the gantry and looking north towards the mountains and home, although the sky was still dark grey, the visibility was extraordinary. I began to search for and mark the places I'd been on my journey around the bay, across to Sunderland Point, though I had to work hard to find it, and on to Heysham Head and further to Silverdale and to where the hills began. Although the great rolling hills of Furness were visible, the coast was tucked in behind the point at Rampside where the bay swung north. Over Piel Island sunlight and shadows worked, illuminating the few buildings and the castle like stage lighting. This was some kingdom to reign over, and I was glad that a woman had made it hers – even for just a few decades.

I said my farewells to the school group and to Brian and walked out to the west coast and along the final curved beach where the outfall of the bay ran, the place I'd seen the little terns for the first time. But in October it was an empty landscape, hardly a bird in sight. Bordering the beach the marram grass of the dunes bristled in great gusts of wind. The shingle beach had been remade in waves and crests by recent storms. A boat came out from the channel and increased its speed as it set off towards the windfarm. I sat in the shelter of the dunes, looking east.

★ ★

Back home again I read more about Peggy and her island domain; of her careering down the island road in her Triumph Herald, roof down, and standing up to shoot rabbits with her .22 rifle. She'd been taught to shoot by her father. His party trick for visitors to the lighthouse was to spin coins into the air, take aim and blast them to smithereens. In the Second World War there had been a small prisoner of war camp on the island and young men, Italian and German, some as young as 15, lived there. They had been brought to the island to help with the job of growing food for British army trainees. The prisoners lived in poverty, possessing little if anything and had little in the way of comfort. But Peggy and other islanders took the boys under their wings. Most of the prisoners didn't even possess shoes. Peggy's family provided them with socks and shoes and knitted bobble hats for them and gave them clothes. One Christmas morning the men invited Peggy's family over to their hut. The prisoners gathered together in a rough ensemble and began to sing to them the carol, *Silent Night*. It was, they said, all they had to give.

**

While researching Walney lighthouse I stumbled upon a short, black and white Pathé news-reel film made in 1948. Remarkably, it was about another female lightkeeper, albeit assistant keeper, Mrs Parkinson, who lived and worked just across the bay at Plover Scar lighthouse beside the outflow of the River Lune, and just south of where my journey began. Plover Scar is a small lighthouse in comparison to most. It's a fancy that could have been imagined into existence by an illustrator of children's books. Because of the aged, monochrome medium, the film seems to have an odd, closed-world atmosphere. In the opening sequence there's a short panoramic of the bay and just visible in the background, the Furness hills. Mrs Parkinson strides over the still-wet sands in knee-high wellington boots and along a causeway to the lighthouse. She climbs the vertical metal-framed ladder up the outside of the tower, and enters the watch-room. She polishes the reflectors, fuels the lamp with paraffin and sets it alight with a match. Underneath her hat, her hair is styled in the wartime Victory Roll, coiled and pinned into place like a never-breaking wave.

The strange thing was that both Peggy and Mrs Parkinson shared facial characteristics. Both were quite square of jaw and were resolute-looking women, women who seemed to have an intrinsic sense of self. Although their lifespan and their lighthouse service didn't completely overlap, their work at either extremity of Morecambe Bay made them guardians of its waters and inner sanctum. I like to think of them both, these remarkable women, polishing their lenses, cleaning the glass at the top of their towers, and occasionally sending greetings to each other across the sea. Brief messages painted on sweeping beams of light.

★ ★

Towards the end of my journey around the bay I had heard the news of my father's very sudden death. In the way of these things, I was far away on a Hebridean island and with limited mobile phone coverage. Seeing the holiday cottage owner walking up the path towards the front door I'd had an immediate sense that something was out of kilter.

'Your brother's trying to get in touch. Come up to the house if you need to use the phone.'

A few minutes later we were talking. With the lack of phone reception, after phoning our neighbour at home and the police on Stornoway, Andy had resorted to calling holiday cottages; he got us on the third call.

'It's Dad,' he said. 'He died this morning.'

So, it had happened.

As I listened to my brother talking, taking in the details of my father's last moments, I looked out of the wide window to the chimneys of the next house where two crows were travelling in backwards in the strong wind, landing on the pots like Harrier Jump Jets in reverse, over and over.

My reaction, after taking stock, was of what a ridiculous place to be at a time like that. That day we'd been out by boat to explore the abandoned island of Mingulay. Later though, I reasoned that it had been an entirely appropriate place. Mingulay was an island that had faced its own end of life, the inhabitants giving up through expediency and necessity, much like St Kilda, and I'd read that they too had found it hard to settle, to accept the change of place and circumstance. They'd built new houses facing south on the sea-washed turf of Vatersay, and from here they could look back and see their former home, just a hump of grey rock in the distance. What had been only a shift of a small number of miles, to them was seismic.

* *

After my visit to the lighthouse I set off to drive up the island road but then stopped again to look across the channel towards Barrow. The massive submarine sheds of BAE Systems had been constructed after my father left his job at the shipyard and moved away again with my mother in the late '70s. They had chosen to leave the town beside the bay and return to my mother's native Merseyside, the place she was never properly able to leave behind, and in pursuit of some illusory idyll. Needless to say, it was never found.

I thought of my father's work at the shipyard, and of how serendipitous it had been – his redundancy and then the new job in Barrow and out of that my coming to live in this landscape and learning to love it all. But actually there was more to it than that; even though I hadn't been born here, what I had from this land was a feeling of being grounded.

On the day of receiving the news of Dad's death I had been doing what I had chosen to do, to venture out, to look and to keep on looking; to explore new ground from the firm foundations of my own territory.

The actual place of my birth now seems an irrelevance; I feel no sense of connection with it of any kind. Perhaps just a few memories from early childhood that occasionally filter through in filmic, grainy images: my grandparents' house where we had all lived; something resembling grief as I stared at the cracked map of a strange ceiling from my cot, waiting interminably for my parents to come home from their only ever trip away – to Paris; dressing up in the garden, strings of beads down to my knees.

Although in my young adult years I did feel something akin to loss, or even fraud that I didn't quite 'belong' *because* of an accident of birth, this feeling was later replaced by the opposite, of an innate sense of belonging; I feel that I am shot through with the truth of being Cumbrian. And here on the shore of the bay, was my ground, and my journey around its circumference turned out to be a metaphor for finding and losing, for connecting and disconnecting, for remembering and forgetting.

The land, *this* land, is ultimately the thing that has sustained me. And not to find joy, or myself, in this landscape, is something I couldn't begin to imagine.

Notes

1 T.W. Potter and R.D. Andrews, 'Excavation and Survey at St Patrick's Chapel and St Peter's Church, Heysham, Lancashire, 1977–8', *The Antiquaries Journal*, Volume 74, Issue 1, March 1994

2 J.D. Bu'lock, *The Pre-Norman Churches of Old Heysham*, Lancashire and Cheshire Antiquarian Society, 1974 (referred to in T.W. Potter and R.D. Andrews article, above)

3 Mark Edmonds, *The Langdales: Landscape & Prehistory in a Lakeland Valley*, The History Press, 2004

4 Francis Pryor, *Britain BC*, HarperCollins, 2004

5 Mark Edmonds, *The Langdales: Landscape & Prehistory in a Lakeland Valley*, The History Press, 2004

6 W.A. Cummins, 'The Neolithic Stone Axe Trade in Britain', *Antiquity*, Volume 48, Issue 191, September 1974 (referred to in: Francis Pryor, *Britain BC*, HarperCollins, 2004)

7 Roy Adkins and Ralph Jackson, *Neolithic Stone and Flint Axes from the River Thames*, Department of Prehistoric and Romano-British Antiquities, British Museum, 1978 (referred to in: Francis Pryor, *Britain BC*, HarperCollins, 2004)

8 Dr Patricia Crown, 'The Golden Age of British Watercolor in the 18th and 19th Centuries', *The Newsletter of The Watercolor USA Honor Society*, Fall 2005

9 Richard Dorment, 'David Cox Exhibition – Review', *The Telegraph*, 16th March 2009

10 John Dennis quoted in: Robert Macfarlane, *Mountains of the Mind*, Granta, 2003

11 Rebecca Solnit, *A Book of Migrations*, Verso Books, 1997

12 Dylan Thomas, 'The force that through the green fuse drives the flower', 1934

13 Elizabeth Gaskell quoted in: Harold Orel (editor), *The Brontës: Interviews and Recollections*, Palgrave Macmillan, 1996

14 D. Hodgkinson, E. Huckerby, R. Middleton and C.E. Wells (editors), *The Lowland Wetlands of Cumbria* (North West Wetlands Survey 6), Lancaster Imprints, 2000

15 J.A. Barnes, 'Ancient Corduroy Roads near Gilpin Bridge', *Transactions of the Cumberland and Westmorland Antiquarian & Archaeological Society*, Volume 4, 1904

16 *Ibid*

17 Ian R. Smith, David M. Wilkinson and Hannah J. O'Regan, 'New Lateglacial fauna and early Mesolithic human remains from northern England', *Journal of Quaternary Science*, Volume 28, Issue 6, August 2013

18 John Bolton, 'On the Kirkhead Cave, near Ulverston', *Journal of the Royal Anthropological Society of London*, Volume 2, 1864

19 John Bolton, *Geological Fragments Collected Principally From Rambles Among The Rocks of Furness and Cartmel*, D. Atkinson and Whittaker & Co, 1869

20 Jim Crumley, *The Last Wolf*, Birlinn, 2010.

21 Aubrey Burl, *The Stone Circles of Britain, Ireland, and Brittany*, Yale University Press, 2000

22 'There is a land of the living and a land of the dead and the bridge is love, the only survival, the only meaning.' Thornton Wilder, *The Bridge of San Luis Rey*, Boni & Liveright, 1927

23 J. Richard Eiser, Joop Pligt and Russell Spears, *Nuclear Neighbourhoods: Community Responses to Reactor Siting*, University of Exeter Press, 1995

24 See the Ship Inn's website: www.pielisland.co.uk

25 Martin Wainwright, 'Peggy Braithwaite – Obituary', *The Guardian*, 20th January 1996

26 Walney Light is maintained by Lancaster Port Authority. It was built as a second navigation light for shipping entering the River Lune, along with Plover Scar Light at Cockersands.

Acknowledgements

Thanks are due to the following people: first and foremost to Kathleen Jamie, without whose diligent expertise and guidance this book would not have taken flight, and also to Meaghan Delahunt and Liam Murray Bell of the Creative Writing Department at Stirling University. To Sara Hunt at Saraband for her utter faith in the book, and to Craig Hillsley for editorial expertise. To the late Cedric Robinson for his generosity in sharing the bay with me, and to his two assistants, Andy Mortimer and Barry Keelan. I am grateful to Thor Ewing for details on the Viking Hogsback Stone. Andrew Davies of Lancashire Archaeology Service. Martin Forwood and Janine Allis-Smith of Cumbrians Opposed to a Radioactive Environment. Pete Moser. Victoria Eden. Hsiao-Hung Pai. Michelle Cooper at the Lancaster Maritime Museum. Staff at Newton Stewart Library. Brian Fereday and John Dunbavin for their in-depth understanding of Foulshaw Moss. Stephen Read of Levens History group for guiding me towards the archaeology of Cumbria's wetlands. Alan Sledmore and fellow guide and Morecambe Bay fisherman Stephen Clarke. Dave Coward for talking me through the details of the Kents Bank and Kirkhead excavations. To Brian Hardwick and Astrid Specht. Robin Horner and other members of staff at RSPB Leighton Moss. Dr George Agidas, Department of Engineering, Lancaster University. B.J. Samuels, Harbourmaster for Lancaster Port Authority, for facilitating privileged access to Walney Lighthouse.

John Murphy, guide to Piel Sands. Keith and Karen Coulthard. The publication *Crossing Lancaster Sands*, produced by Heysham Heritage Association, was invaluable for early research. To Laura Mitchell for early reading of chapters. To Steve, Callum and Fergus for the ongoing gifts of space, time and support during and well beyond my year of being away. There are countless people who I have spoken to or met whilst walking the edgelands and during research. Thanks to all those. Any omission is entirely my own.

...and not forgetting Harold Wilson.

KAREN LLOYD is a writer based in Cumbria and currently the Writer in Residence with Lancaster University's Future Places Centre. Her most recent book, *Abundance: Nature in Recovery* (Bloomsbury, 2021) was longlisted for the 2022 James Cropper Wainwright Prize for Writing on Conservation. *The Gathering Tide* is her first book; it was followed by *The Blackbird Diaries: A Year with Wildlife* (Saraband, 2018), and both won Lakeland Book Awards. She is the editor of *North Country: An anthology of landscape and nature* (Saraband, 2022). Her poems have appeared in a number of anthologies. Karen produced and edited the *Curlew Curling* anthology (2017) to raise awareness and funds for curlew conservation. She holds a PhD in Creative Writing from Lancaster University, where she also teaches.